SCIENCE 1

Fourth Edition

bju press
Greenville, South Carolina

SCIENCE 1
Fourth Edition

Coordinating Writer
Leah R. Thomas, MEd

Writers
Rosanna Capellan, MS
Debra Harrold White

Biblical Worldview
Bryan Smith, PhD
Tyler Trometer, MDiv

Academic Oversight
Jeff Heath, EdD

Project Editors
Kristin Kowalk, MA
Joanna Lynch, MA

Consultant
L. Michelle Rosier

Concept Designers
Josh Frederick
Emily Heinz

Designer
Emily Heinz

Page Layout
Faith Bupe Mazunda

Cover Designer
Emily Heinz

Cover Illustrator
Dana Thompson

Art & Design Facilitator
Jim Frasier

Illustrators
Emily Heinz
Jack Pullan
Dana Thompson

Project Coordinators
Tony Every
Jennifer Ferguson

Permissions
Sharon Belknap
Sylvia Gass
Carrie Hanna
Ashleigh Schieber
Kathleen Thompson

Photo credits appear on page 262.

Greenville, South Carolina 29609

978-1-62856-313-9

15 14 13 12 11 10 9 8 7

Contents

Let's Learn About Science 1

Chapter 1

Big Question

Why did God give people five senses?

Science and Scientists

God created our world.
He created all that is in it.
Last of all, God created people.

God gave people minds.
He gave us minds to learn about His world.
God created people to be curious.
God created us to ask questions.
What questions do you ask about God's world?

Lesson 2

What Is Science?

God's world is full of amazing things.

Those things are all around you.

Science is the study of God's world.

God wants you to learn about His world.

God wants you to ask questions about His world.

Science can help you answer questions.

What questions do you have about God's world?

Science can answer some questions.
What do you add to the dirt to grow plants?
What happens to ice in the sun?
Science can answer these questions.

Science cannot answer some questions.
Where did the world come from?
Why are you here?
Science cannot answer these questions.

What is science?

Can this frog swim?
Can science answer your question?

God tells us to love our neighbors.
People study science to show love.
People study science to help others.
You can learn what a plant needs.
This helps you grow a flower.
You can give it to a neighbor who is sick.
You are using science to help others.

God tells us to care for the earth.
People study science to care for the earth.
You can learn what an animal needs.
This helps you care for animals.
You can care for dogs at a shelter.
You are using science to care for the earth.

Why is science important?

EXPLORATION

Process Skill

- Observe

Looking at God's World

In this Exploration you will look at God's world.

Your home is part of God's world.

Everything around you is part of God's world.

Let's take a closer look at your home.

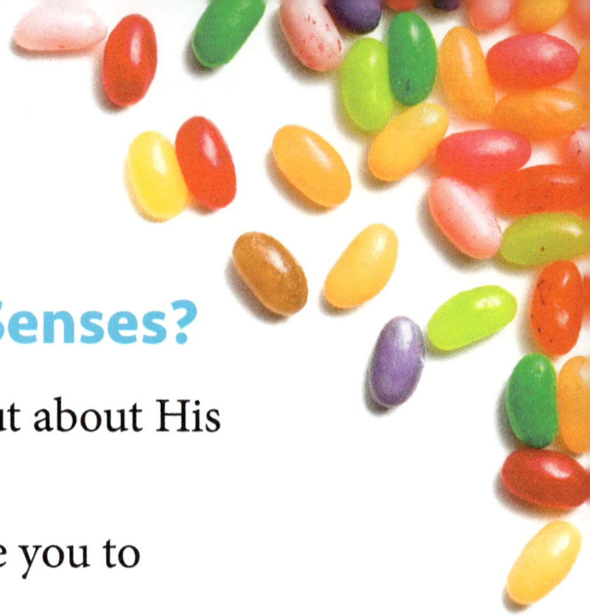

What Are the Five Senses?

God gave you ways to find out about His world.

Senses are the ways God gave you to find out about the world.

Your five senses are sight, touch, smell, taste, and hearing.

Sight

You see with your eyes.

You can see shapes or sizes of things.

You see many different things.

Your sense of sight helps you find out about God's world.

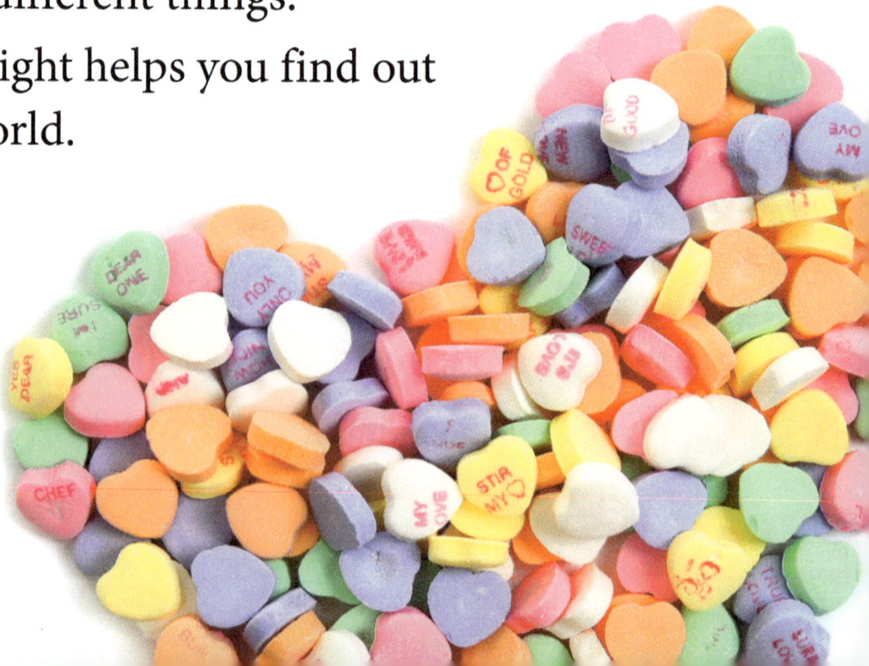

Touch

You use your hands and fingers to touch.

You touch bumpy or smooth things.

You touch many different things.

Your sense of touch helps you find out about God's world.

Smell

You use your nose to smell.

You can discover good smells or stinky smells.

You smell many different things.

Your sense of smell helps you find out about God's world.

What are the five senses?

Taste

You use your tongue to taste.
You taste things that are sweet.
You taste things that are sour.
You taste many different things.
Your sense of taste helps you
find out about God's world.

Hearing

You use your ears to hear.

You hear loud things and soft things.

You hear bangs and whistles.

You hear many different things.

Your sense of hearing helps you find out about God's world.

What do you use to taste?

God gave you many ways to find out about His world.

You can use your senses.

You can observe.

To **observe** means to use your senses to find out something.

You observe to find out about God's world.

What does the word *observe* mean?

Lesson 4

What Is a Scientist?

A **scientist** is a person who studies God's world.

Scientists use their senses to observe the world.

Scientists are curious.

They ask questions.

They try to find answers to help others.

God made the world.
The world is God's world.
He tells people to care for the earth.
He tells us to love our neighbors.
A scientist shows love when he helps other people.

What does a scientist do?

STEM Careers

Engineer

An **engineer** uses science and math to find answers to problems.
Engineers **design**, or plan, something.
They use a design process.
The design process has steps.

There are different types of engineers.
Some engineers design cars.
Others design buildings.
Some engineers design robots.

Do you like to find answers to problems?
Maybe you can be an engineer.

STEM

The Engineering Design Process

What Is a Scientist's Worldview?

You study science to learn about God's world.
Scientists study God's world too.
The way a person thinks about the world is called a **worldview**.
All scientists have a worldview.
A scientist's worldview helps him.
It helps him explain what he finds.
Not all scientists believe what the Bible teaches.
Let's take a look at a biblical worldview of science.

Where did our world come from?

God created the earth.

God created the land and plants.

God created the sun, moon, and stars.

God created the animals.

He created man and woman.

God created all things in six days.

God created a perfect world.

Why do things in our world work together the way they do?

God is the Engineer of the earth.

He designed the earth.

God's design is beautiful.

God is the Artist of the earth.

He designed the parts of the earth
to work together.

Each part works with the other parts.

They work together like gears in a machine.

How is God an engineer?

Who are we? Why are we here?

People are the most important part of God's creation.

In Genesis 1:26 God said, "Let us make man in our image, after our likeness."

God created you in His **image**, or likeness.

God created all people in His image.

You are like God in many ways.

You are like God through your work.

God tells people to rule over the earth.

God tells you to care for the earth.

God tells you to help others.

Why is science important?

Science is helpful in your life.

You use science to see God's glory.

The world shows you that God is great.

Science helps you learn how nature works.

Then you can use nature to help people.

You use science to care for the earth.

How can you use nature?

Meet the Scientist

Sir Francis Bacon was a scientist from long ago. He believed in God and the Bible. He found a way to study science. He started with facts. He used tests to answer problems. His plan is known as the scientific method.

Chapter 2

Big Question

How can we learn about God's world?

What Scientists Do

Scientists study God's world.
Scientists use skills to help them.
They also use tools to help them.
They do tests to find answers to problems.
You can be like a scientist.
You can learn about God's world.

Lesson 9

Science Process Skills

Scientists use different skills to study the world.

These are called science process skills.

A **science process skill** helps a scientist discover and use facts.

You can use science process skills to learn about God's world.

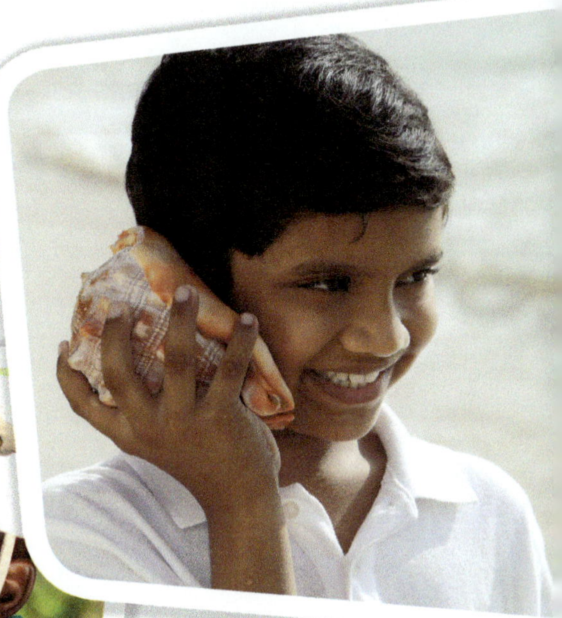

Observe

You observe to learn about God's world.

You see, hear, taste, smell, and touch.

You use your senses to observe things.

Why do we use science process skills?

Classify

To **classify** means to group things that are alike.

You can group things by kind or size.

You can classify shoes by kind or size.

You can group things by color or shape.

You can classify blocks by color or shape.

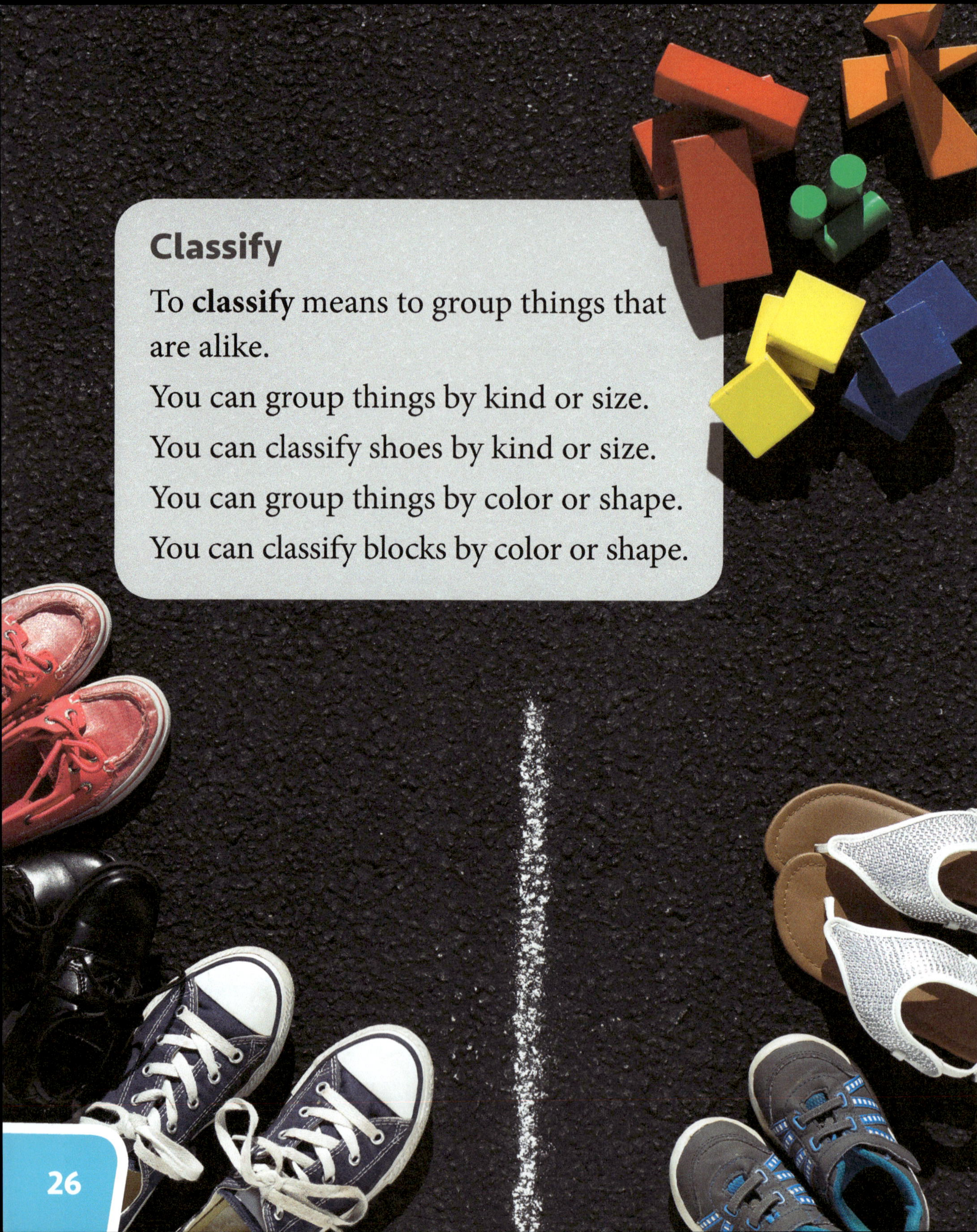

Measure

To **measure** means to find the size, number, or amount of something.

You can measure how long a pencil is.

You can find out the number of crayons in a box.

You can find out how heavy a box is.

You can find out how much water to give a plant.

Why do we measure?

Infer

To infer means to use what you know to tell why things happen.

You infer after you observe.

You know that ice is cold.

You infer what ice will do to water.

Is the water hot or cold?

Predict

To predict means to make a careful guess at what may happen.

You predict after you observe.

You observe that the sun's light is warm.

You can predict what ice will do in the sun's light.

What does it mean to predict?

Communicate

You communicate when you share what you know.

You may tell others what you know.

You may show others what you know.

Science Tools

A tool makes work easier.

A **science tool** is an object that helps scientists work.

They use tools to find out about God's world.

Tools help them observe and measure.

You can use science tools to observe and measure.

Why do we use science tools?

Hand Lens

You use a hand lens to observe small things.
A hand lens makes things look bigger.
Moving the lens makes things look larger.
Moving the lens makes things look clearer.

Ruler

You use a ruler to see how long something is.

Some rulers measure in inches.

Some rulers measure in centimeters.

Some rulers measure in both.

Scientists measure in centimeters.

What does a hand lens help us observe?

Measuring Cup

You use a measuring cup to measure how much space something takes up.

Beakers are measuring cups that scientists use.

Beakers measure in milliliters.

Balance Scale

You use a balance scale to measure the weight of something.

A balance scale weighs objects to find out how heavy they are.

A balance scale weighs objects in grams or pounds.

What does a balance scale measure?

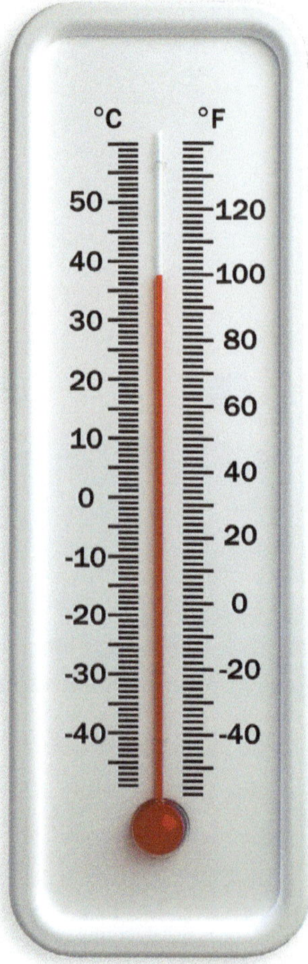

Thermometer

You use a thermometer to measure temperature. **Temperature** tells how hot or cold something is. Most scientists measure in degrees Celsius.

What is measured with a thermometer?

EXPLORATION

Process Skills

- Observe
- Measure

Lesson 12

Using Science Tools

You can explore God's world.

Science tools can help you explore.

Science tools help you observe and measure.

In this Exploration you will practice using science tools.

A Science Investigation

Scientists study God's world.
They try to find answers to questions.
They try to solve problems.
They use investigations to do these things.

An **investigation** is a test to find an answer to a question.
An investigation follows steps.
The steps are part of the *scientific method.*
You can follow these steps to do an investigation.

Problem

An investigation begins with a problem. First decide what you want to know. Then write a question about it. You ask a question if you want an answer to a problem.

What question do you want to answer?

Hypothesis

Make a hypothesis. A **hypothesis** is a possible answer to the question.

What is an investigation?

Materials

Gather the materials.

The materials are the supplies needed to do the investigation.

Procedure

The procedure is the steps of the investigation.

Follow the steps carefully to test your hypothesis.

Observations

Observe what happens.

Use your five senses to observe.

Write down what you observe.

Draw what you observe.

Conclusions

Read your hypothesis.

Study your observations.

Do your observations match your hypothesis?

Draw conclusions to tell what you learned.

Conclusions are the answer to your test.

What are the conclusions in a science investigation?

STEM

Lesson 14

How to Keep My Pencil on My Desk

An engineer finds answers to everyday problems.

When he answers problems he is obeying God.

In Genesis 1:28 God tells people to rule over the earth.

An engineer uses math and science to rule over the earth.

An engineer follows a plan with steps.

The engineering design process is the plan an engineer follows.

In this STEM activity you will find an answer to an everyday problem.

You will use the engineering design process.

Let's Learn About **2**

Living Things

Chapter 3

Big Question

How are plants designed by God to survive and grow?

Plants

God made our world.

God made plants on the third day of Creation.

God made plants for people to use.

God designed plants to grow.

God designed plants to make the same kind of plants.

Lesson 17

Living Things and Nonliving Things

God made living things in our world.

A **living** thing must have food, water, air, and space to live.

Living things grow and change.

God made you.

You are living.

Can you think of other living things God made?

God also made nonliving things.

A **nonliving** thing does not have to have food, water, and air.

Nonliving things do not grow or change.

God made light.

God made rocks and dirt.

God made water and air.

These are all nonliving things.

How are living things different from nonliving things?

Needs of Plants

Plants are living things made by God.
All living things have needs.
A **need** is what a living thing must have to stay alive.
A plant has five main needs to live and grow.

Light and Air

Two of a plant's needs are *light* and *air.*
The plant's leaves use light and air to make food.
Food gives energy.
Energy is what is needed to cause change.
Energy from the sun's light helps a plant grow and change.

Water, Dirt, and Space

Water and *dirt* help a plant live and grow.

The plant gets water that it needs from the dirt.

Some plants need lots of water.

Some plants need only a little water to live and grow.

The last need of a plant is *space*.

A plant needs space to grow larger.

Some plants need lots of space to live and grow.

Some plants need only a little space.

What are the five needs of plants?

Ways People Use Plants

God created plants for people to use.
God made different kinds of plants.
Some plants are used for food.
Other plants are used for clothing.
Some plants are used to build houses.

Result of Sin

God created plants.

They were perfect.

The Bible says in Genesis 3:17–18 that plants changed.

Plants changed when Adam sinned.

Plants are no longer perfect.

There are now weeds, thorns, and thistles.

Weeds take up space and make it hard for other plants to grow.

Thorns and thistles make some plants hard for people to use.

What happened to plants after Adam sinned?

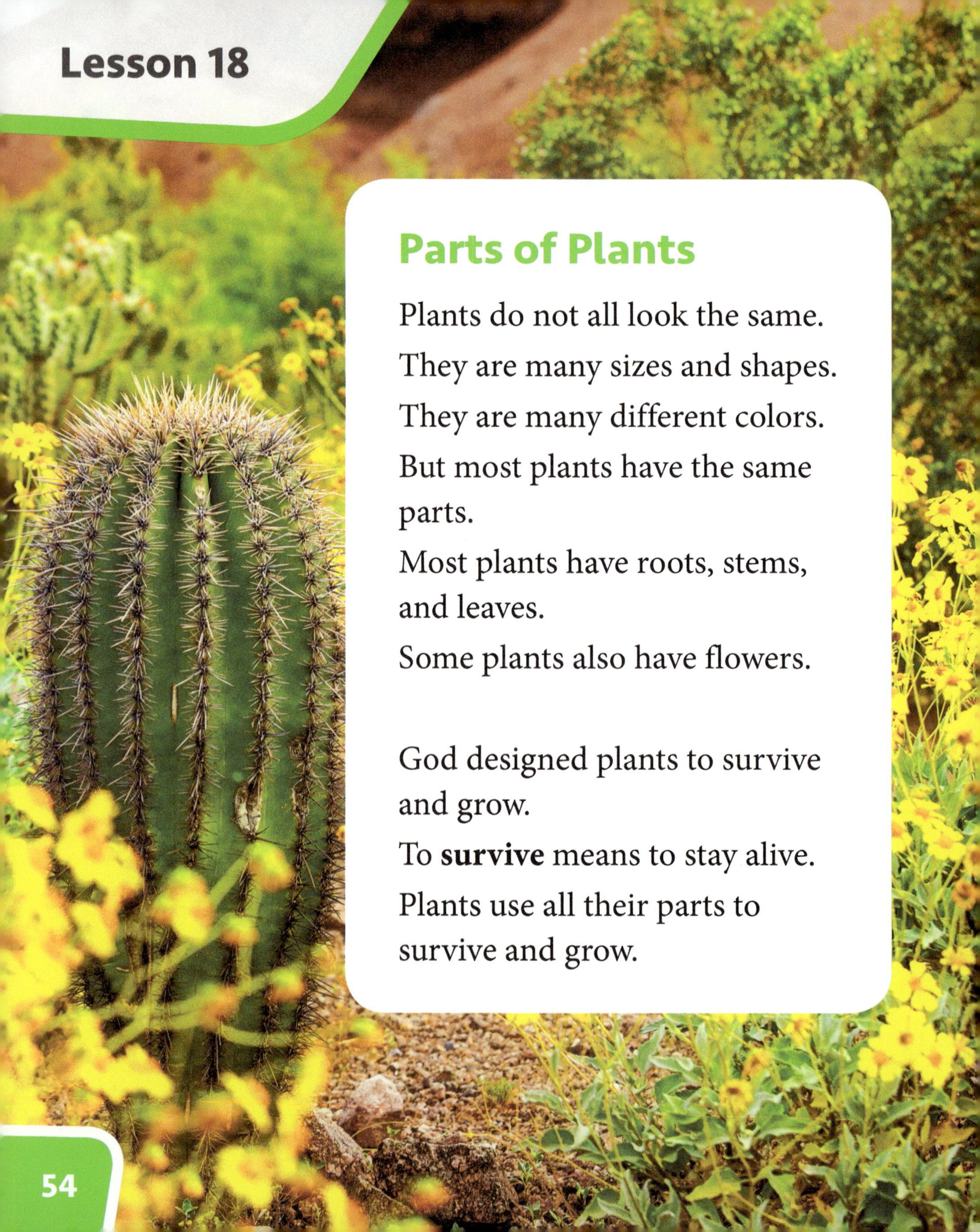

Lesson 18

Parts of Plants

Plants do not all look the same.
They are many sizes and shapes.
They are many different colors.
But most plants have the same parts.
Most plants have roots, stems, and leaves.
Some plants also have flowers.

God designed plants to survive and grow.
To **survive** means to stay alive.
Plants use all their parts to survive and grow.

What are the four parts of a plant?

Roots

Roots hold the plant in the ground and take in water.

Roots are under the ground.

The roots take in water from the ground.

Roots help the plant survive and grow.

What roots do people eat?

What stems do people eat?

Stems

The **stem** holds up the plant.

Water from the roots moves up the stem.

A tree has a large stem.

A dandelion has a small stem.

The stem helps the plant survive and grow.

What is the job of the roots?

There are many shapes and sizes of leaves.

Leaves

Plants make their own food.

Leaves make food for the plant.

Leaves use light, air, and water to make food.

Leaves help the plant survive and grow.

What leaves do people eat?

Flowers

Many plants have flowers.

The **flower** makes seeds.

A new plant will grow from a seed.

Flowers also add color and beauty to plants.

God designed all parts of the plant.

God designed the plant parts to work together.

Plants survive and grow because the roots, stem, leaves, and flowers work together.

What flowers do people eat?

What is the flower's job?

INVESTIGATION

Lesson 19

Plant Needs

Process Skills
- Predict
- Observe

All living things have needs.

What do plants need to survive and grow?

In this Investigation you will predict what will happen to a plant that does not get water.

You will make a hypothesis.

You will observe to check your hypothesis.

Let's investigate!

A Plant's Life Cycle

Living things grow and change.
All the steps of the life of a living thing are a **life cycle**.

A plant's life cycle starts with a *seed*.
Most new plants grow from seeds.
Plant a seed and a seedling will grow.

A *seedling* is a young plant.
The seedling has roots.
It also has a thin stem and small leaves.

The seedling grows into an *adult plant*.

The plant grows flowers.

The flowers grow fruit.

A pumpkin is fruit.

There are seeds inside the fruit.

The seeds will make new plants.

Life Cycle of a Pumpkin Plant

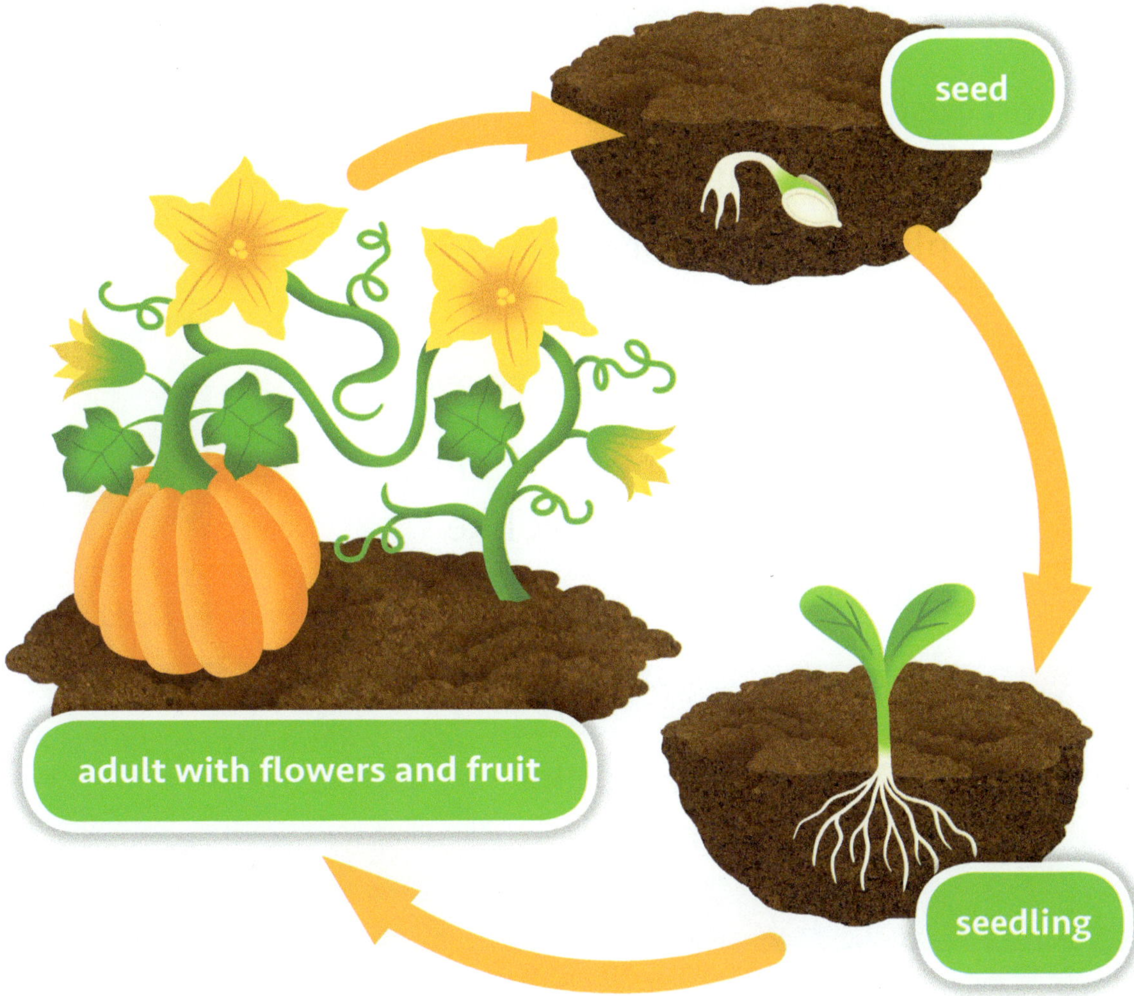

What are the steps of a plant's life cycle?

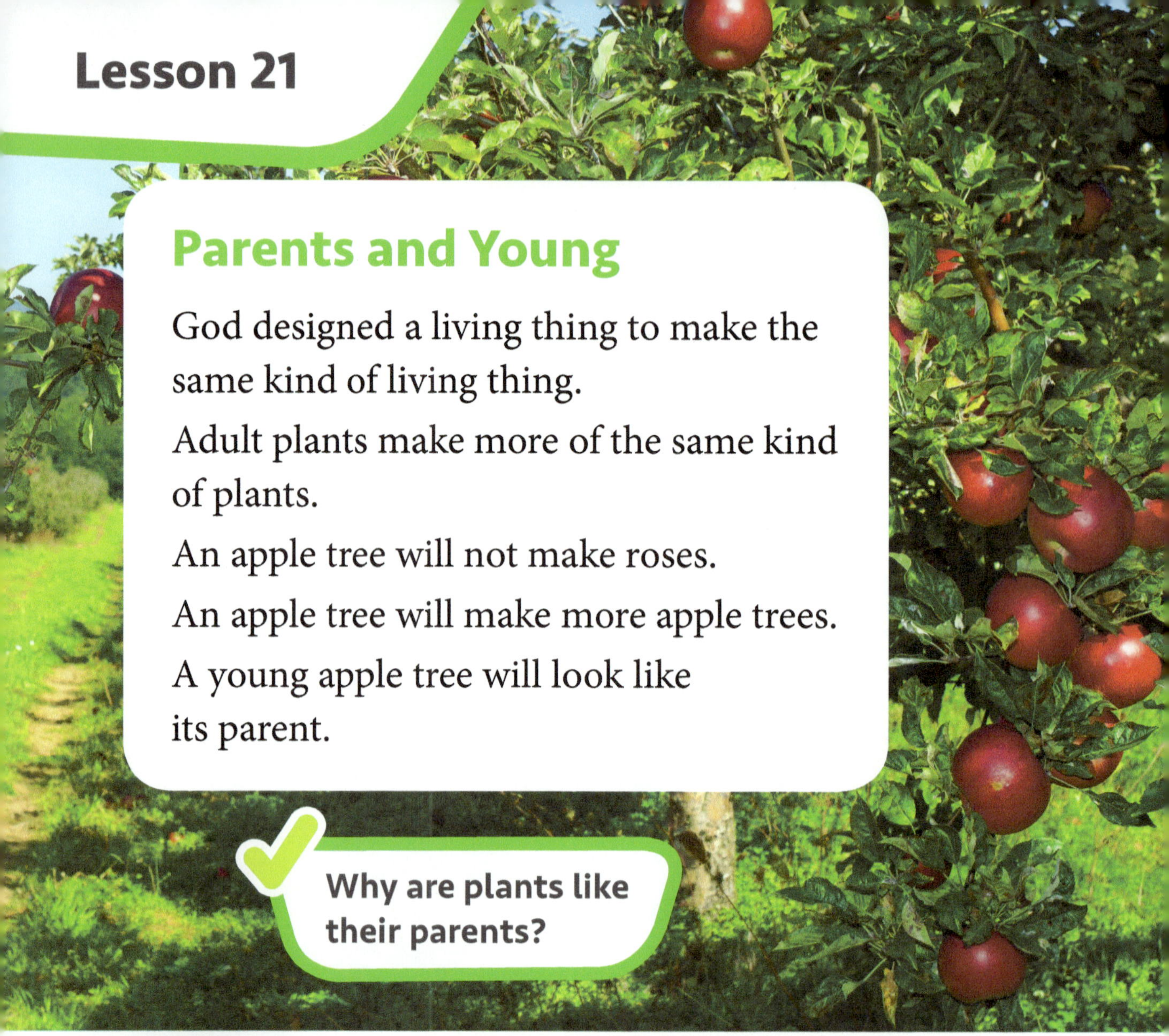

Parents and Young

God designed a living thing to make the same kind of living thing.

Adult plants make more of the same kind of plants.

An apple tree will not make roses.

An apple tree will make more apple trees.

A young apple tree will look like its parent.

Why are plants like their parents?

Meet the Scientist

As a child George Washington Carver was called the plant doctor. He learned to care for the family garden. George was curious. He enjoyed studying plants. He gave advice about plants to his friends. He helped many sick plants survive and grow.

STEM

Lesson 22

Unwanted Plants

There are weeds growing in your garden.

The weeds are taking up space that your other plants need to grow.

You do not want the weeds to grow there.

You want to remove the weeds.

You do not want to hurt the other plants.

How can you get rid of the weeds that are growing in your garden?

Chapter 4

Big Question

How are animals designed by God to survive and grow?

Animals

God created all things.

He created nonliving things.

The air and rocks are nonliving things.

God created living things.

Plants and animals are living things.

In Proverbs the Bible tells people to take care of animals.

We give glory to God when we obey Him.

Environment

Imagine you and a friend are walking in a park.

You see flowers and grass.

Birds fly overhead.

Insects zoom by.

You hop across a creek on rocks.

You see a rabbit sitting very still.

The park is one environment.

An **environment** is all the living and nonliving things around a plant or animal.

What living things are in a park?

What nonliving things are in a park?

Can you find living and nonliving things in the park?

What is an environment?

Needs of Animals

God designed animals.

An animal has five main needs to live.

Its needs can be met in its environment.

God designed an animal's environment so that the animal can survive and grow.

Air, Water, and Food

Three of an animal's needs are *air*, *water*, and *food*.

An animal breathes air and needs water to live.

Food gives energy and helps an animal grow.

Air, water, and food are found in an animal's environment.

Does the frog have its needs met?

Shelter and Space

A *shelter* is a safe place to live.
A cow may have a barn.
A bug may use a plant for shelter.
A fox has a den.
All animals need shelter.

A fifth need of an animal is *space* to grow.
Animals need space to move and find food.
They need space to care for their young.
Shelter and space are found in an animal's environment.
God designed animals and their environments to work together so that the animals can survive and grow.

What five needs must animals have?

Classifying Animals

Adam named all the animals.

We do not know the names he used.

Scientists have names for animals too.

These names help us classify animals.

Scientists look at the parts of animals.

They use these parts to classify animals.

Fish

A **fish** lives in water and has gills.

A fish spends all its life in water.

Birds

macaw parrot

A **bird** is an animal with feathers and wings.

Feathers help a bird stay warm.

Feathers can also help a bird fly.

God gave birds two wings.

Some birds use their wings to fly.

The ostrich is the largest bird.

It has wings, but it does not fly.

How is a bird different from a fish?

robins

ostrich

An ostrich runs fast.

Mammals

Most pets are mammals.

And most farm animals are mammals.

A **mammal** is an animal with hair or fur.

Many mammals have hair or fur all over their bodies.

Dogs and lions have fur all over their bodies.

Some mammals only have small amounts of hair on their bodies.

Elephants and whales have small amounts of hair.

What kind of animal has fur or hair?

STEM Careers

Zoologist

A **zoologist** is a scientist.
A zoologist learns how to care for animals.
He finds out about animal environments.
He learns what animals need to grow and survive.

Zoologists work in different places.
Some observe animals in their own environment.
Some work in aquariums or zoos.
Some teach in classrooms.
Zoologists enjoy keeping animals healthy and safe.

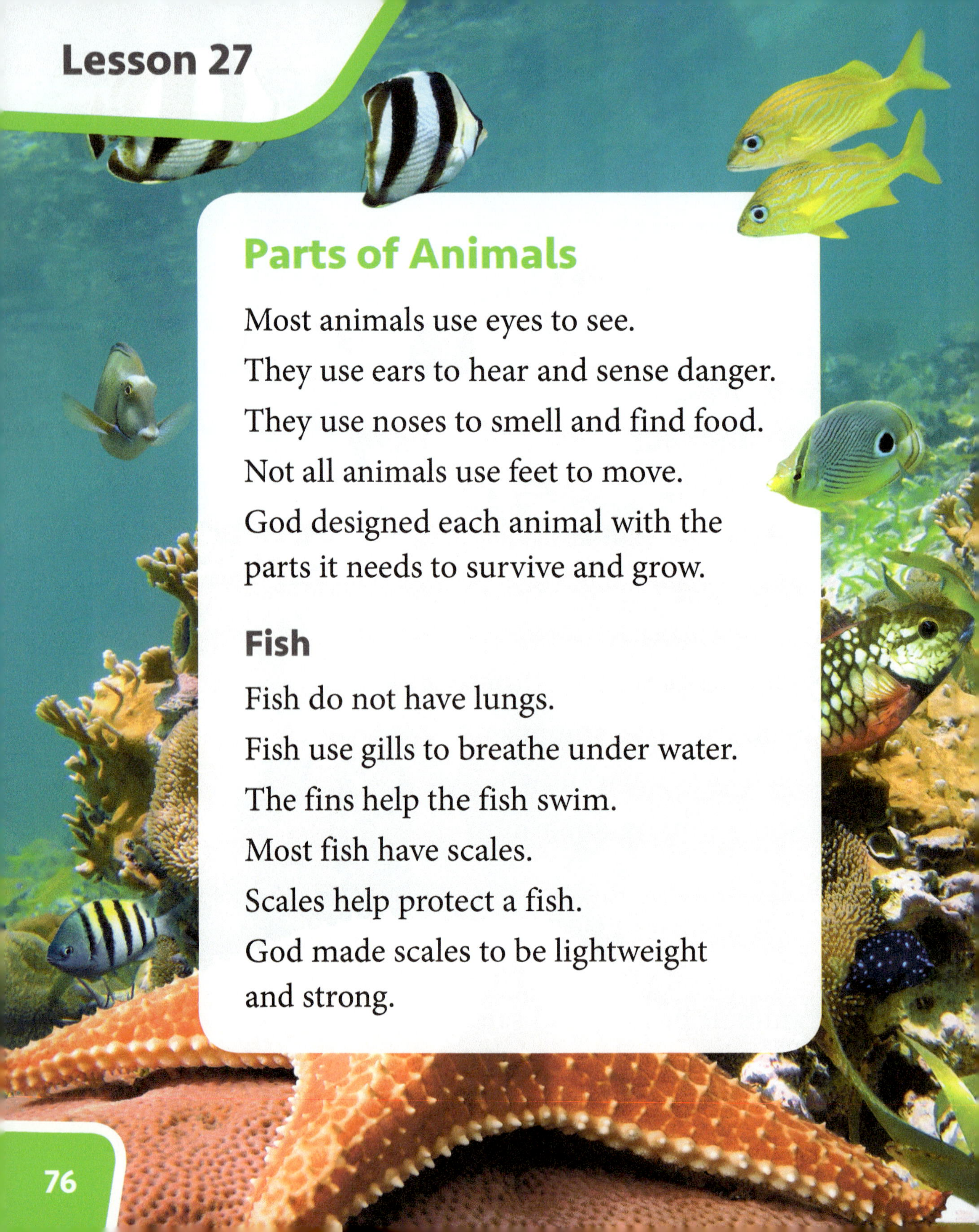

Parts of Animals

Most animals use eyes to see.
They use ears to hear and sense danger.
They use noses to smell and find food.
Not all animals use feet to move.
God designed each animal with the parts it needs to survive and grow.

Fish

Fish do not have lungs.
Fish use gills to breathe under water.
The fins help the fish swim.
Most fish have scales.
Scales help protect a fish.
God made scales to be lightweight and strong.

Birds

All birds breathe with lungs.

Most birds fly with their wings to find food.

Some birds use their beaks to get food.

Some birds use their feet to catch food.

Many birds carry twigs in their beaks.

They use twigs to make nests.

An eagle catches its food with its feet.

This Vitelline masked weaver is making a shelter for its babies.

What protects a fish?

Mammals

Some mammals have hooves.

A hoof is the hard part of a mammal's foot.

Hooves help protect the animal's feet.

Some mammals use their front legs as hands.

Their front legs can hold things.

Fantastic Facts

The trunk on an elephant helps the elephant smell and touch. It also brings food and water to an elephant's mouth. A mother elephant hugs her baby with her trunk. Elephants use their trunks to splash each other at play.

Some mammals have long tails.
They use their tails like another arm or hand.
Front legs or tails can help them get food.
Some mammals use just their mouths for getting food.
Mammals have different kinds of shelters.
Monkeys have trees as their shelters.

God designed birds with the right body parts to survive.
He designed the fish and mammals to be able to get food.
God designed each animal with the parts it needs to survive and grow.

What can mammals use to help them get food?

A Robin's Life Cycle

Some animals lay eggs.

Some animals give birth to live babies.

All the steps of the life of an animal are called a life cycle.

Most animals have a simple life cycle.

Their life cycle has three steps: eggs, young, and adult.

Eggs

A mother robin builds her nest.

She lays three to five eggs.

She sits on her eggs to keep them warm.

Young

A robin's young are called chicks.
The parents feed the chicks fruit and worms.
The young will grow feathers and learn to fly.

Adult

The young will grow and change.
An adult can find its own food.
An adult will continue the life cycle.

What are the three steps in a robin's life cycle?

Parents and Their Offspring

Many **offspring,** or babies, need care.
Parents build shelters to keep them safe.
Parents keep their offspring warm.
Offspring cry or make a cheeping sound.
Parents know to take offspring food.
Parents move their babies to keep them safe.
They carry offspring in different ways.
Some offspring do not need care after birth.

Baby sea turtles know how to find their food.

How Parents and Offspring Are Alike

God said each animal will make the same kind of animal.

The offspring will be the same kind as its parents.

A dog will have puppies.

A dog does not have kittens.

The puppy and the parents are alike.

The puppy has the same parts as its parents.

What does God say about an animal's offspring?

How Parents and Offspring Are Different

Offspring and their parents can be different.

A baby is not exactly like its parents.

The offspring is smaller than its parents.

A baby can be a different color than its parents.

A baby may have different spots than its parents.

A Dalmatian puppy does not have its spots yet.

A zebra has different stripes than its parents.

Result of Sin

God created the animals.

He created them perfect.

He created animals to eat plants.

When Adam sinned, animals changed.

Now some animals eat other animals.

This is called death.

Some animals also die because their needs are not met.

Some animals die because they are sick or old.

Death is now a part of the life cycle.

Death happened because of Adam's sin, or the Fall.

What caused animals to die?

STEM

Lesson 29

Copying God's Design

Plants have parts to help them survive and grow.

Plants have roots, stems, leaves, and flowers.

Animals have parts to help them survive and grow.

Most animals have eyes, a nose, ears, and legs.

Some animals have long necks and hard shells.

People get ideas from these parts.
They use the ideas to design something that will help solve a problem.
How can an animal body part help solve a human problem?
How can a plant part help solve a human problem?

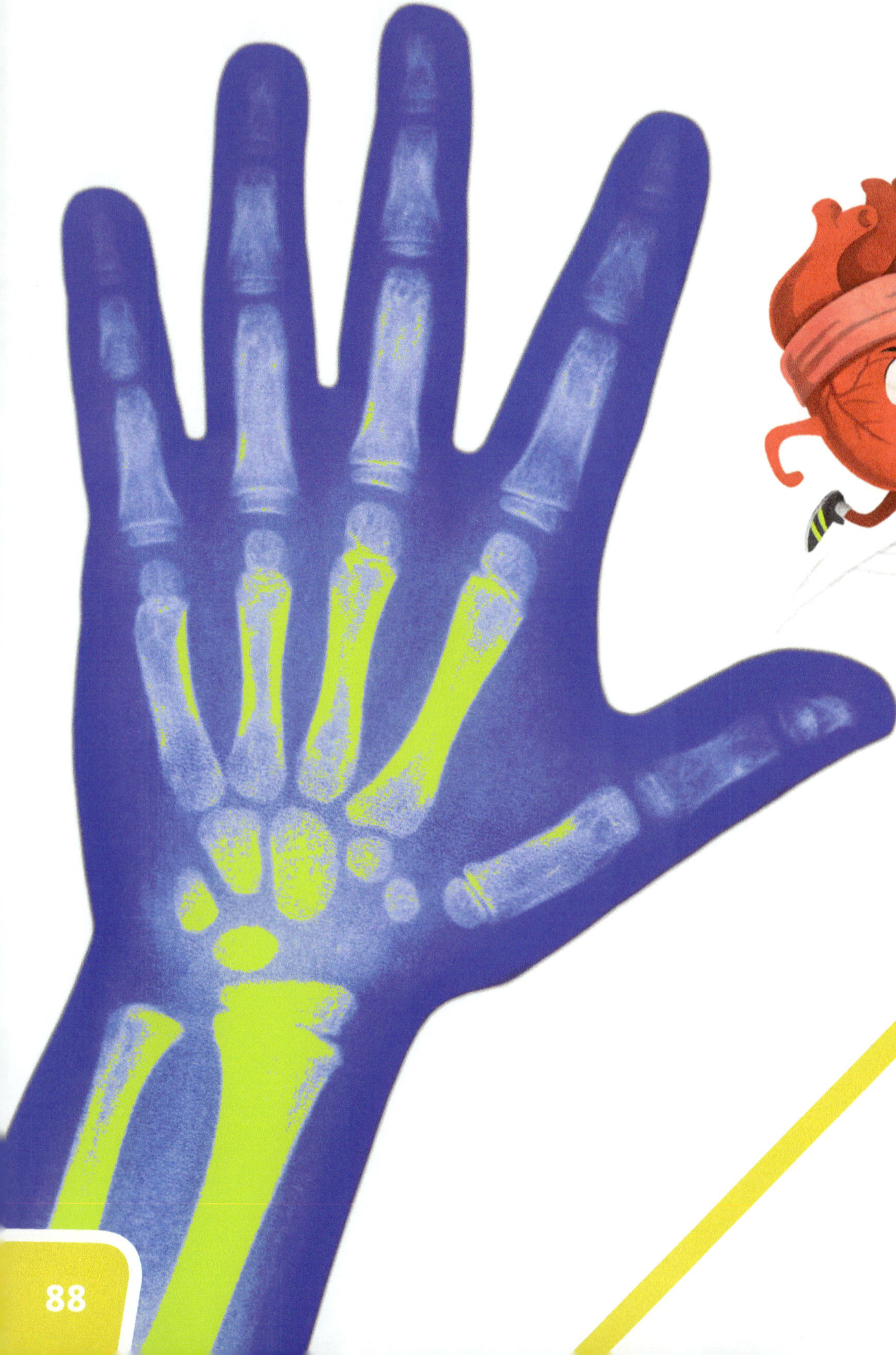

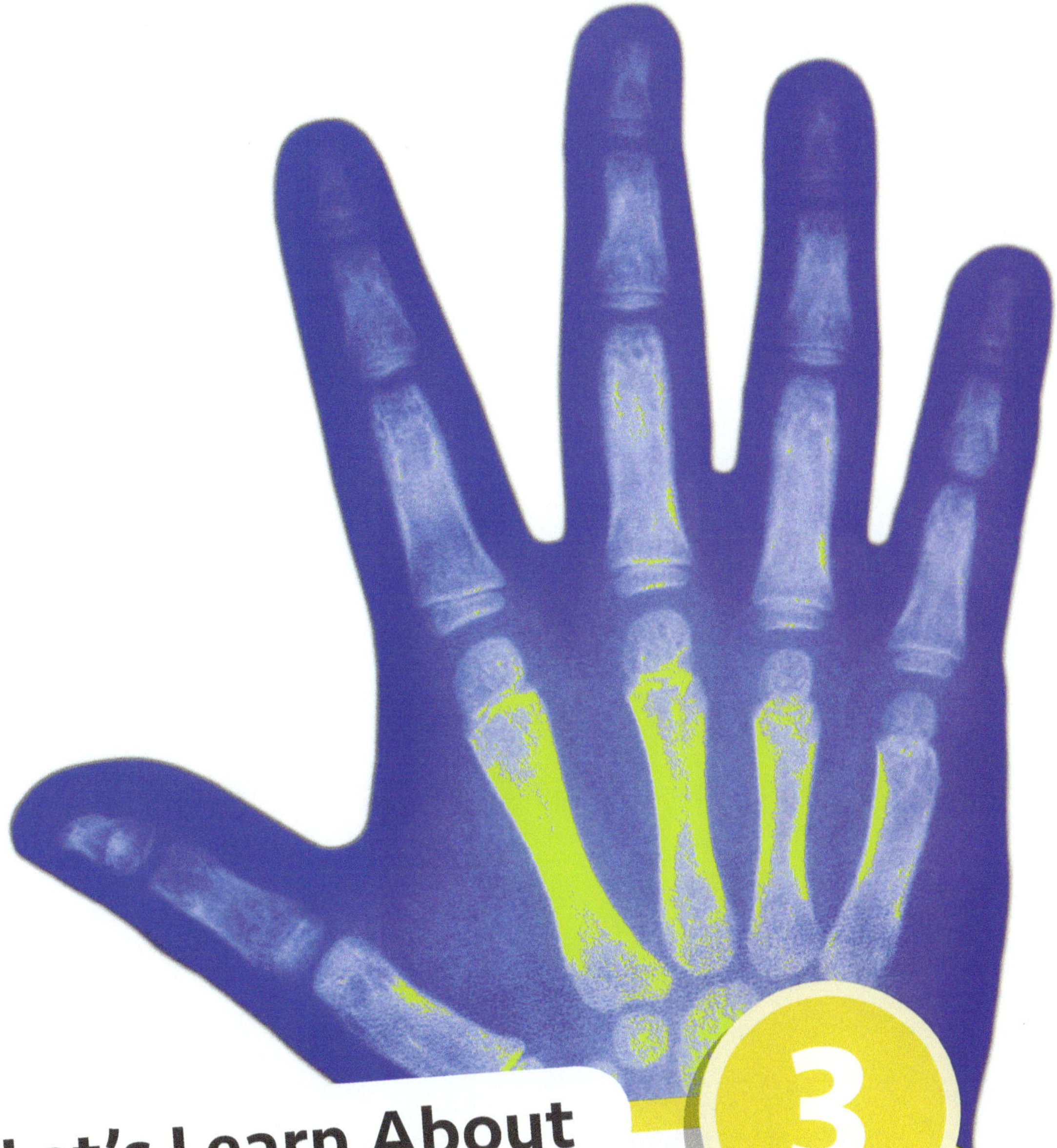

Let's Learn About

3

Our Bodies

Chapter 5

Big Question

How does God's design of my body help me survive and grow?

The Human Body

Plants and animals are living things.

People are alive too.

All living things have needs.

Living things meet their needs in their environments.

God designed living things and their environments.

He designed them to work together so that living things can survive and grow.

Lesson 32

Animals and People

How We Are the Same

Animals are living things.
People are alive too.
Both have the same needs.
Both need air, water, food, shelter, and space.
All their needs are met in their environments.
God designed people, animals, and their environments to work together so they can survive and grow.

How We Are Different

The Bible tells us that God created people different from animals.

God created the first man, Adam, from the dust of the ground.

God formed the first woman from Adam's rib.

He created all people in His image.

God did not create animals in His image.

People are created in God's image.

Therefore we can talk with God and each other.

God can talk with people through the Bible.

People can love God and each other.

All people will live somewhere forever.

Each person God created is different.

Each person is special to God.

We can say with the psalmist, "I will praise thee; for I am fearfully and wonderfully made" (Psalm 139:14).

How are people different from animals?

EXPLORATION

Process Skills
- Observe
- Infer

Lesson 33

My Head

God created you just the way He wanted you to be.

We should praise God for making each of us special.

To **praise** means to thank God for who He is and what He does.

In this Exploration, you will study your head.

You will name the parts of your head.

You will tell God's purpose for each part.

How are your five senses used?

How will you praise God with a part?

Outside Your Body

Your Head

God designed your head.

He placed it at the top of your body.

The **head** holds the brain.

Your head has its own parts to help you survive and grow.

God gave you *eyes* to see His creation.

You see the food you eat.

You see dangers around you.

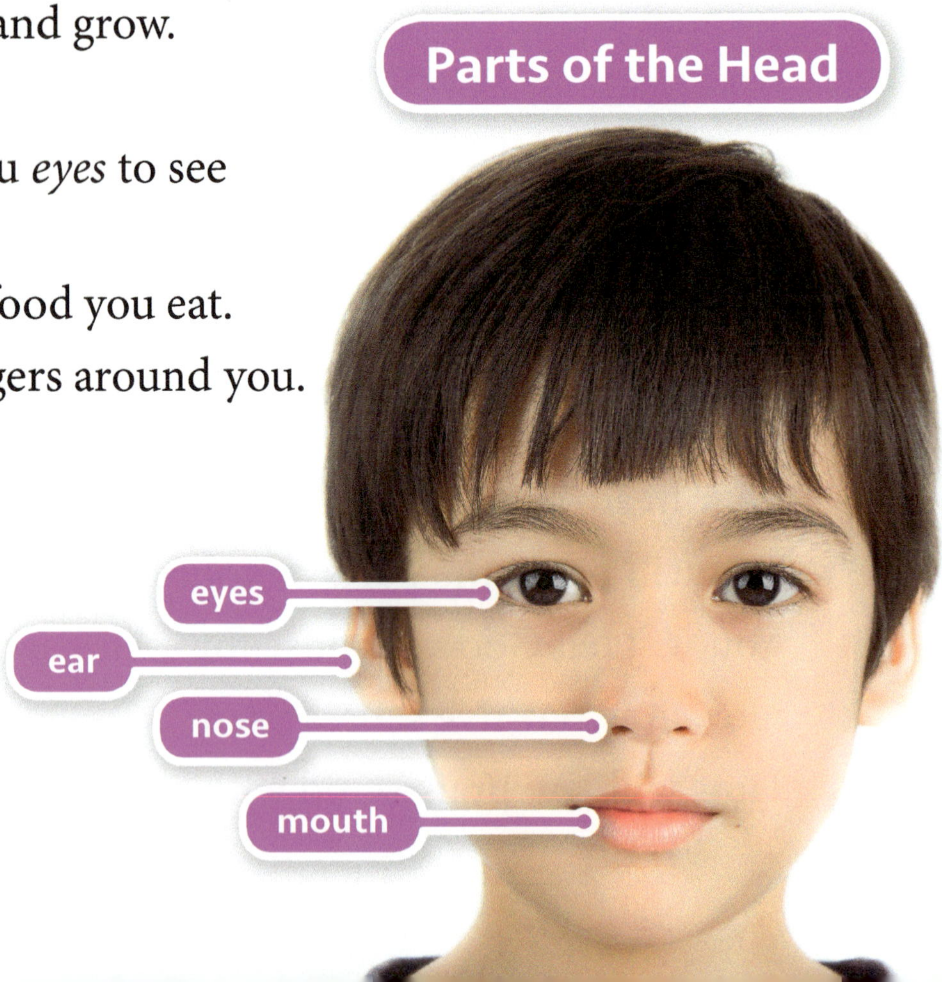

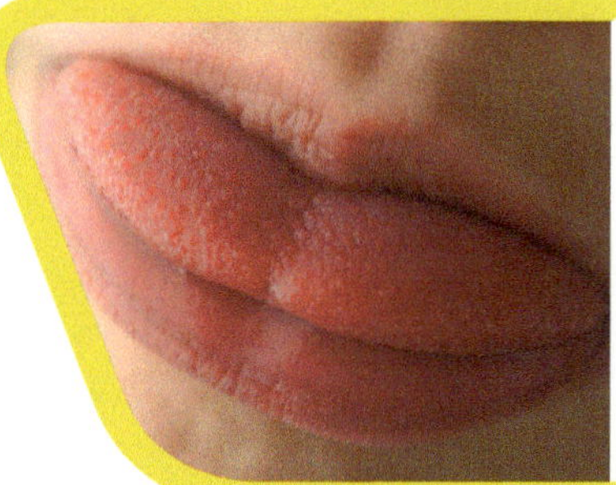

Creation Corner

God created you with a way to taste foods. Is your food sweet or salty? What helps you decide? Your tongue has many taste buds. Taste buds help you taste your food. Taste buds are too small to see, but they have a huge job.

God gave you a *mouth* to speak, sing, and eat.
You talk to family and friends.
You sing and pray to God.

God gave you *ears* to hear.
You hear your mother calling you
to come and eat.
You hear the birds sing.

What did God create at the top of your body?

God gave you a *nose* to smell.
You smell delicious food.
You sniff roses and lilies.
God designed each of these parts so that you can survive and grow.

Your Arms

God designed your arms.
The **arms** let you reach for things.
Your arms bend and move in different ways.
At the end of each arm is a hand.
Your hands have fingers and a thumb.
These help you grab and hold things.

Your Legs

God designed your legs.
The **legs** help you stand, walk, and run.
At the end of each leg is a foot.
Each foot has toes.
Your feet and toes help you keep your balance.
They also help you walk and run.

What are your arms for?

All these parts are on the outside of your body.
These parts help you get food and water.
These parts help you keep safe.
God designed these parts to work together.
He designed them to help you survive
and grow.
Let's praise Him.
We are "fearfully and wonderfully made"
(Psalm 139:14)!

What do the parts on the outside of your body help you do?

Inside Your Body

Your Bones

God designed the inside of your body too.

He designed the bones in your body to work together.

The **bones** are hard parts of the body.

Your bones are alive.

They grow and change.

How did God design your bones?

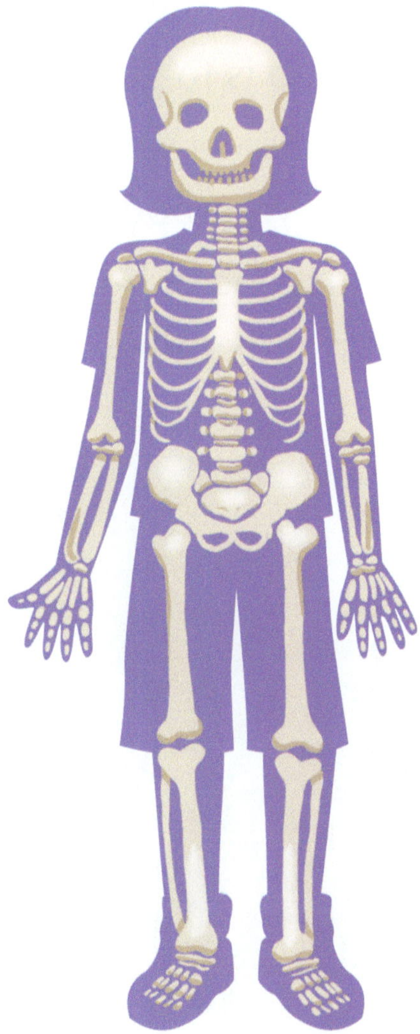

God gave your bones some important jobs.
One job is to give your body shape.
They also protect the soft parts inside your body.
A third job of bones is to help your body move.

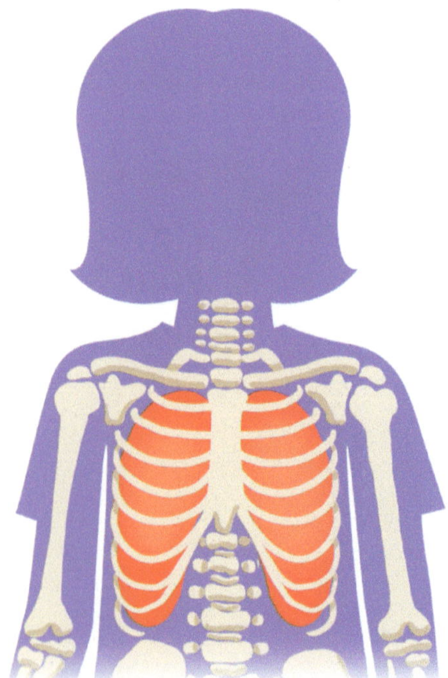

Your Muscles

God designed the muscles to work with the bones.

The **muscles** make the bones and other body parts move.

Many of your muscles connect to bones.

When the muscles move, the bones move.

Some muscles move without you thinking about them.

God made muscles that work even while you sleep.

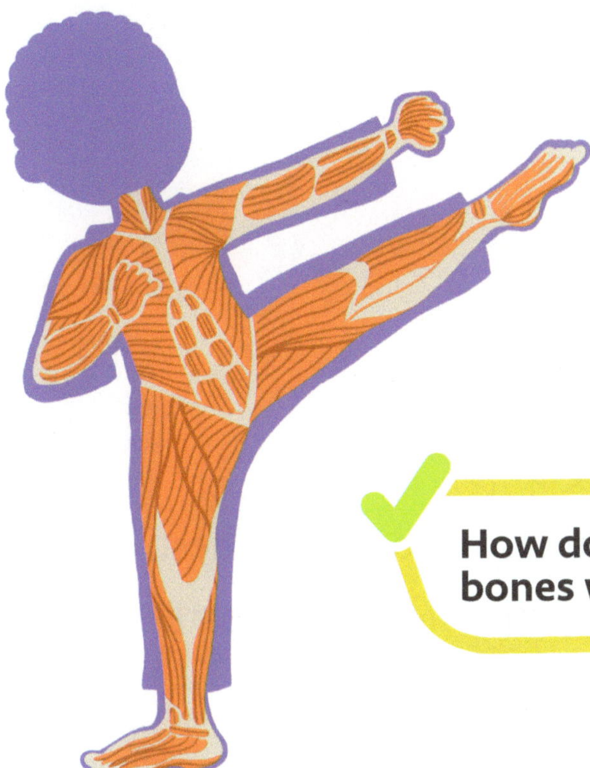

How do the muscles and bones work together?

Your Heart

The **heart** pumps blood through the body.
It is a powerful muscle.
God placed your heart in your chest.
Your heart is about the size of your fist.
Each time it beats, it pumps blood through the body.
It beats all the time.
The only rest your heart gets is between beats.

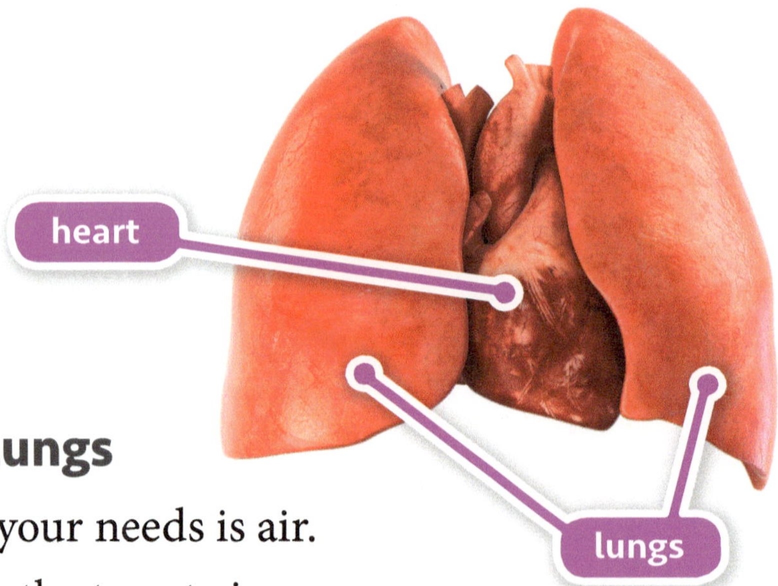

Your Lungs

One of your needs is air.
You breathe to get air.
God placed two lungs inside your chest.
The **lungs** fill with air when you breathe in.

They empty when you breathe out.

You breathe about 20 times every minute.

Genesis 2:7 tells us that God created the first man.

God breathed into him the breath of life.

With this breath, Adam became a living person.

Breathing is a gift from God.

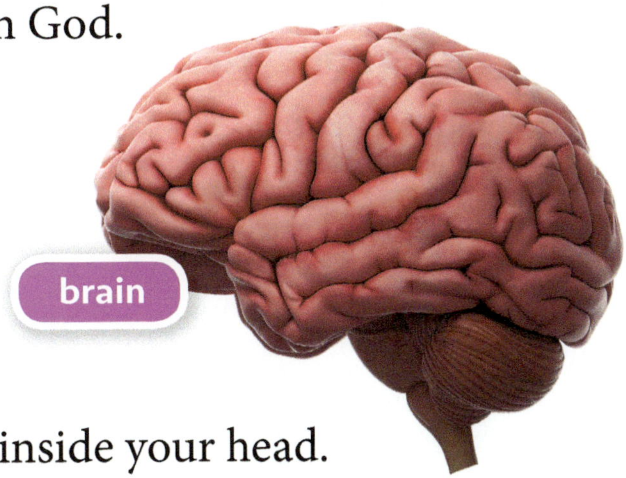

Your Brain

God designed a brain inside your head.

The **brain** controls everything in the body.

It helps you blink, breathe, and see.

It helps you walk, learn, and grow.

The brain never stops working.

What two body parts never stop working?

Your Stomach

You need food to survive and grow.

Your teeth chew food.

Then you swallow the broken-up food.

The food travels down a tube to your stomach.

God designed your stomach.

The **stomach** breaks down the swallowed food.

Your body uses the food from your stomach to survive and grow.

Your body gets energy from the food.

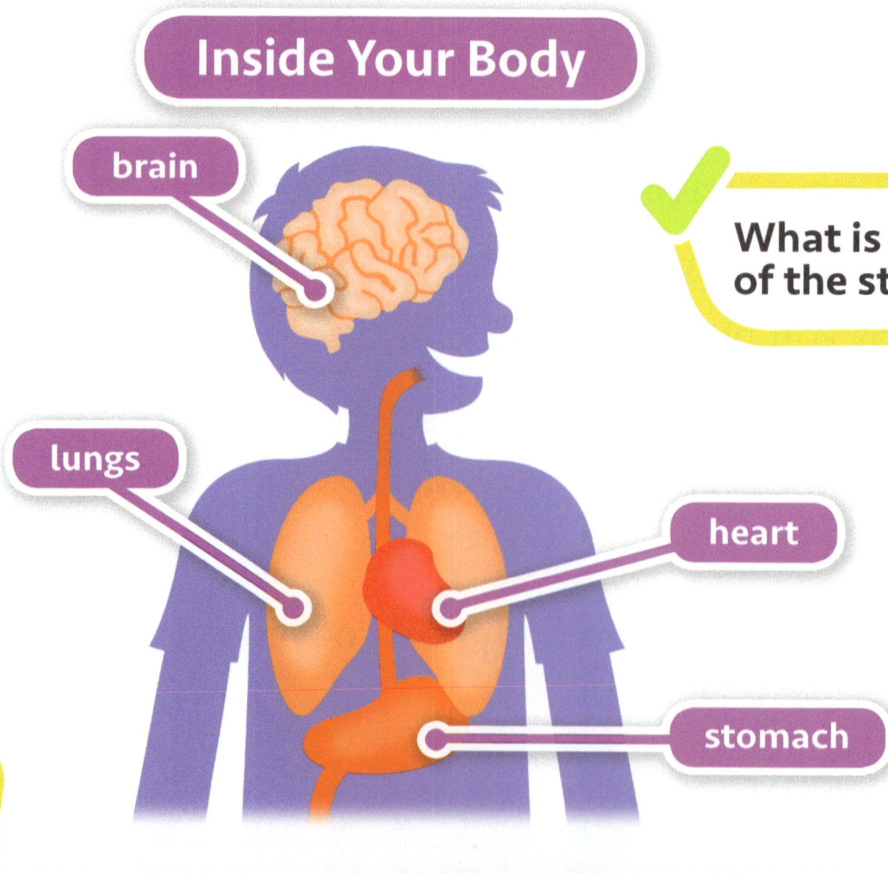

What is the job of the stomach?

EXPLORATION

Process Skills
- Observe
- Communicate

Lesson 36

How My Lungs Work

Close your eyes.

Think about talking, working, or playing.

Did you think about breathing at the same time?

Most of the time, you do not think about breathing.

God created us to breathe without thinking.

How wonderful is God's creation!

In this Exploration, you will show how the lungs work.

You will see how God's design of the lungs helps people survive and grow.

Chapter 6

Big Question

Why are healthy and safe habits important?

Care for the Human Body

God made people in His image.

God made our bodies.

He gave us the job of caring for them.

We glorify God when we take care of our bodies.

We show that we care for other people when we are kind.

Lesson 39

Be Kind

People are the most important part of God's creation.

You are made in God's image.

It pleases God when you are kind to others.

Being kind shows that you care about other people.

In Words

Your words should be kind.

The Bible tells us that kind words are good for other people to hear.

You are kind when you say "thank you."

You are kind when you say "please."

You are kind when you say "hello" to a friend.

In Actions

Your actions should be kind.

You are kind when you smile.

You are kind when you share.

You are kind when you obey.

Kind words and actions show that you love and value other people.

Why should you be kind to others?

Healthy Habits

A **healthy** person is not sick or hurt.

A **habit** is something you do over and over again.

Healthy habits help you care for your body.

For a Strong Body

You should eat food that is good for you.

You should run and play outside.

You should get the sleep you need.

These are healthy habits that will keep your body strong.

What are some healthy habits to keep your body strong?

To Keep Germs Away

Some habits help keep germs away.

Germs are things that can make you sick.

You cannot see germs.

You can get rid of some germs by using soap and water to wash your hands.

You should wash your hands before you eat.

You should use soap and water to keep cuts on your skin clean.

You should take baths and wash your hair to keep clean.

To Keep Germs Away from Others

Some healthy habits help to keep germs away from other people.

You should cover your mouth when you cough or sneeze.

You should use a tissue when you sneeze.

You should throw your used tissue in the trash.

You should wash your hands.

All these habits will help keep others from getting sick.

For Strong Teeth

Brushing your teeth is a healthy habit.

You use your teeth to cut and chew food.

God made your teeth strong.

Brushing helps your teeth stay healthy and strong.

You need to take care of your teeth.

What are three ways to keep germs away from others?

STEM Careers

Dentist

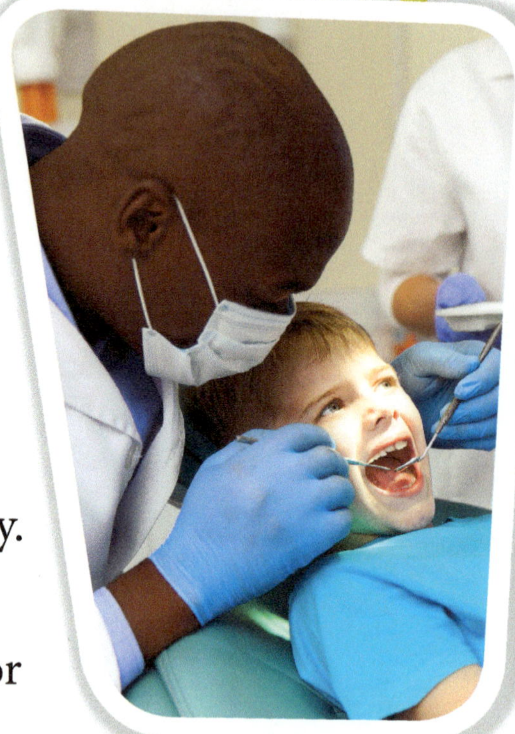

A **dentist** is a doctor who takes care of teeth.

He looks for cavities.

A **cavity** is a hole in your tooth.

A dentist will fill the cavity in your tooth.

He uses special tools to fill the cavity.

A dentist teaches you how to care for your teeth.

Brushing your teeth helps take away food and germs.

Brush your teeth with a toothbrush and toothpaste.

Brush your teeth at least twice a day.

Doing these things helps keep your teeth healthy and strong.

INVESTIGATION

Lesson 41

Clean Hands

Process Skills

- Observe
- Infer

Are your hands clean?

You wash your hands with soap and water.

You are told that this will help get rid of germs.

In this Investigation you will see if washing your hands does help get rid of germs.

Lesson 42

Safe Habits

You should stay away from things that could hurt you.

This is called being **safe**.

Being safe is a way to take care of your body.

At Play

Always tell an adult where you are going.

Wear a helmet when you ride a bike or a scooter.

If it is sunny, wear sunglasses, sunscreen, and a hat.

Obey playground rules.

Doing these things helps keep you safe at play.

In the Car

When in the car, you should not distract the driver.

You should sit in a booster seat.

You should put on your seat belt.

Doing these things helps keep you safe in the car.

What are some safe habits at play and in the car?

Lesson 43

At Home

It is good to have safe habits at home.

Always obey your parents.

Keep your house doors locked.

Do not open your doors to strangers.

Know how to escape if there is a fire.

Know what to do if someone gets sick or hurt.

Fire Safety

1. Be sure your house has a smoke detector.
2. Leave your house quickly if you hear the alarm.
3. Crawl low if there is smoke in a room.
4. Stop, drop, and roll if your hair or clothes catch on fire.
5. Never run back into a burning house.
6. Call 911 when you are safely outside.

In Your Community

It is good to have safe habits when you are away from home.

Stop, look, and listen before you cross a street.

Always have a buddy with you.

Do not talk to strangers.

Do not go with someone you do not know.

Police officers and firefighters are your friends.

Call 911 if there is danger.

What are some safe habits at home and in your community?

God gave you your body.
You should take care of it.
Healthy habits and safe habits are good.
They help you to take care of your body.

Be Safe Around Water

1. Learn to swim.
2. Do not swim alone.
3. Wear a life jacket when on a boat.

STEM

Lesson 44

Safe Shoes

Getting a new pair of shoes can be fun.

Some new shoes have slick soles.

Slick soles can be dangerous.

New shoes can cause a person to slip or fall.

In this STEM activity you will find an answer to slick shoes.

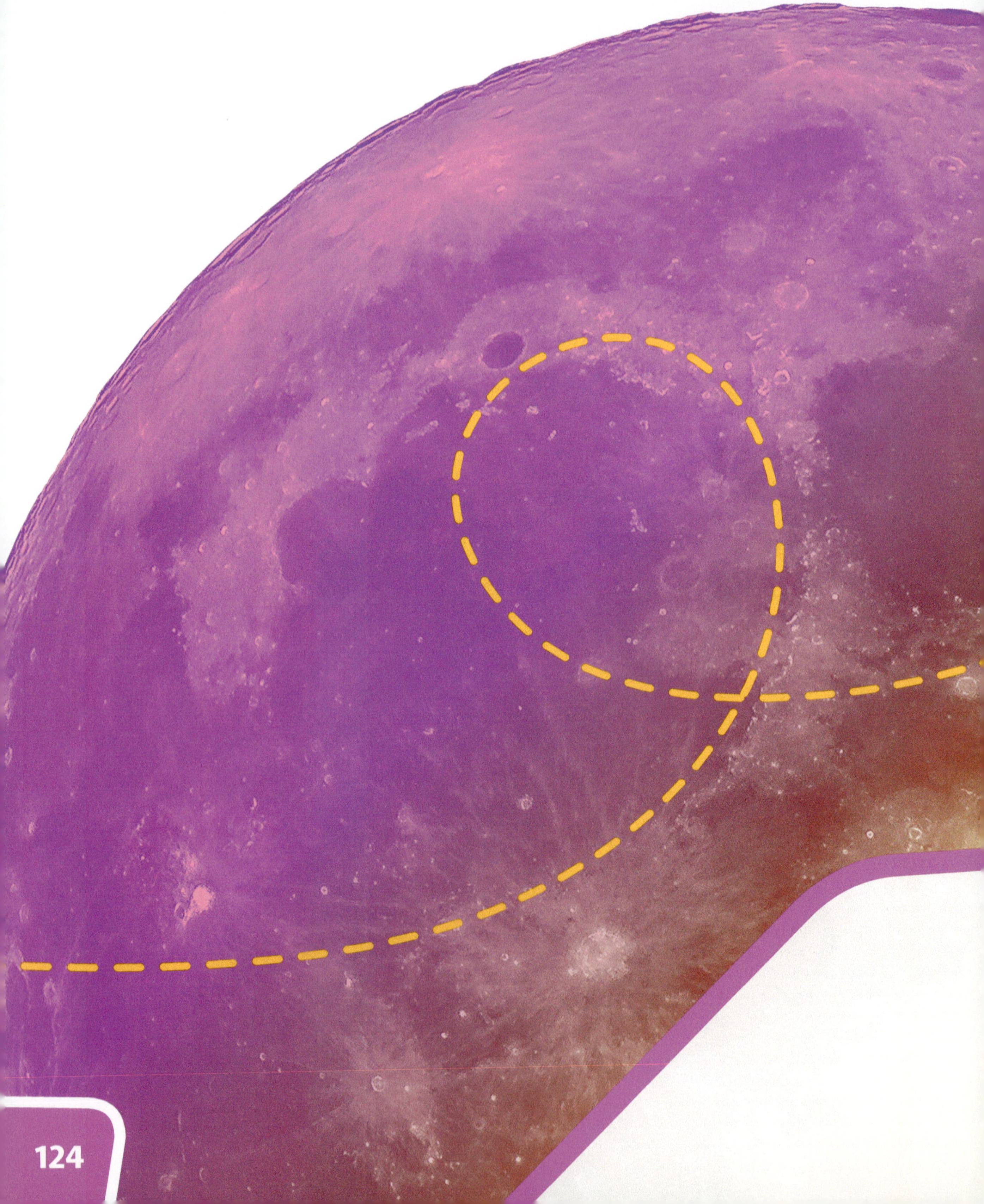

Let's Learn About

4

Earth and Space

Chapter 7

Big Question

How does the sky change each day?

The Earth and Its Lights

The earth was made by God.

When God first made the earth, there was only darkness.

Then God said, “Let there be light.”

He called the light *day*.

He called the darkness *night*.

Later God created the sun, moon, and stars.

God wisely designed the sun, moon, and stars to be lights for the earth.

What the Bible Says

God is great.

God is all-powerful.

The Bible says that God made the earth, sun, moon, and stars.

Genesis 1 tells us that God spoke and the earth was created.

God spoke again and the sun, moon, and stars were made.

God made the earth to be our home.
God made the sun, moon, and stars to be lights in our sky.

How were the earth, sun, moon, and stars made?

What Some Books Say

Some books say that God did not make the earth.

They say that God did not make the sun, moon, and stars.

These books do not agree with the Bible.

When a book does not agree with the Bible, the Bible is right.

The Bible is true because it is God's Word.

How do we know the Bible is true?

The Earth

The earth **rotates**, or spins.

The earth rotates once every day.

A **day** is the time that the earth takes to rotate one time.

We see different lights in the sky because the earth rotates.

What does the earth do once every day?

The Sun

We see the sun in the sky during the daytime.

It is daytime for the part of the earth that faces the sun.

It is nighttime for the part of the earth that does not face the sun.

The sun is very big.

It is bigger than the earth.

The **sun** is a star.

The sun is the closest star to the earth.

A **star** is an object in the sky that makes its own light.

The sun is bright.

The sun is hot.

It warms the earth.

Genesis 1 tells us that God made the sun to give the earth light during the day.

The sun gives light to the earth.

Light is used to see objects around us.

Plants need light to make food.

How does the sun help the earth?

Rotating Earth

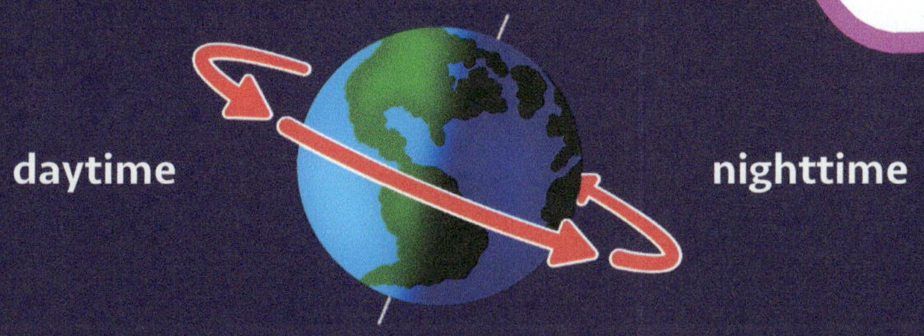

The Path of the Sun

The sun looks like it is moving across the sky.

But the sun is not moving across the sky.

The earth rotates and the sun follows a path.

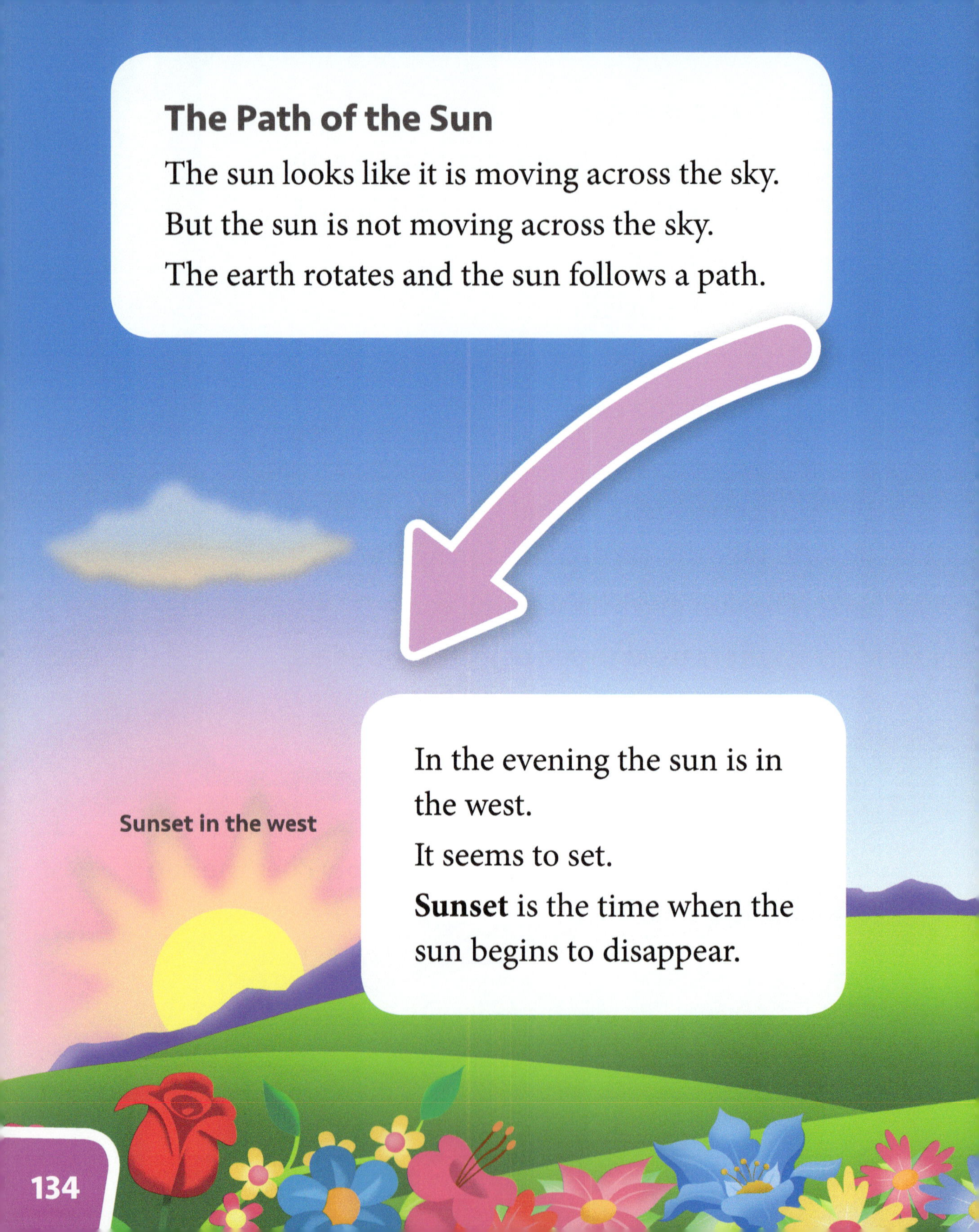

Sunset in the west

In the evening the sun is in the west.

It seems to set.

Sunset is the time when the sun begins to disappear.

The sun is overhead about noon.

In the morning the sun is in the east.

It seems to rise.

Sunrise is the time when we can first see the sun.

Sunrise in the east

Where in the sky will the sun be when you wake up in the morning?

INVESTIGATION

Lesson 49

Stars in the Day

Process Skills
- Observe
- Infer

The sky is full of stars.

Sometimes it is easy to see stars.

Sometimes we cannot see any stars.

In this Investigation you will find out why it is hard to see stars during the daytime.

The Stars

God made many, many stars.
The sun is a star that you see in the daytime.
We see the other stars at night.
The stars are far away from the earth.
They look like points of light.
Like our sun, stars are always shining.
We can see many stars just by looking at the nighttime sky.

But to see even more stars we need a tool.

A **telescope** is a tool that helps us see things that are far away.

A telescope makes objects look closer and bigger.

With a telescope we can see even more stars.

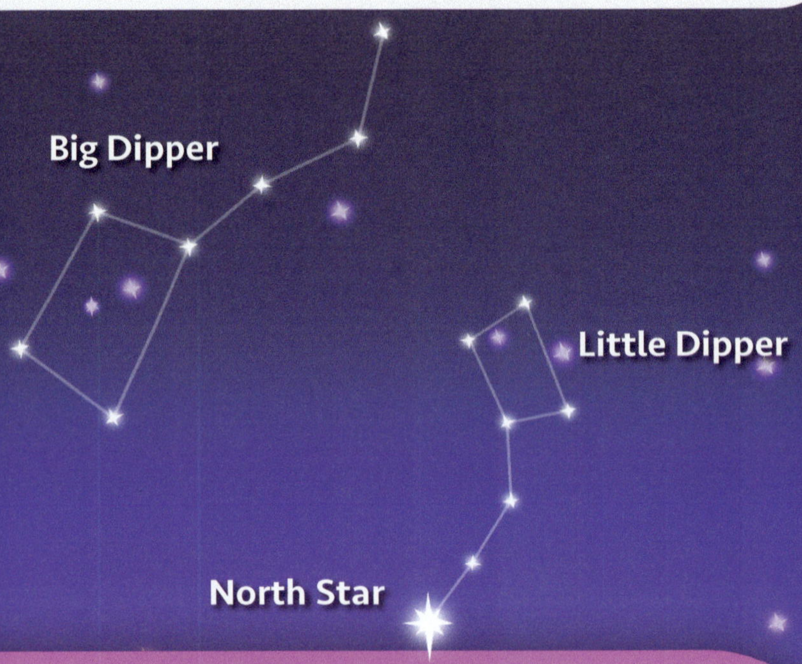

Science and History

Galileo Galilei made his first telescope more than 400 years ago. Galileo observed the sky. He made some important discoveries. He discovered craters on the moon. He discovered Jupiter's four moons. He also discovered spots on the sun's surface. Galileo saw things that no one before had ever seen.

Some groups of stars seem to make pictures.
One group of stars is called the Big Dipper.

One star is bright and easy to see.
It is called the North Star.
The North Star is part of a group of stars called the Little Dipper.

What tool can you use to see more stars?

The Moon

God made the moon to give the earth light at night.

The **moon** is a big ball of rock that moves around the earth.

It does not make its own light.

The moon reflects the light of the sun.

To **reflect** means to bounce off.

Light from the sun bounces off the moon.

The moon is closer to the earth than the sun is.

The moon is smaller than the sun or the earth.

Some people have walked on the moon. These people are called astronauts. An **astronaut** is a person who travels into space.

Why did God make the moon?

Science and History

Alan Shepard was the first American astronaut in space. Shepard went 116 miles above the earth in 1961. His trip into space lasted about 15 minutes. He later led a flight to the moon. He was the fifth man to walk on the moon.

Commander Alan Shepard on the moon

The Path of the Moon

The moon looks like it is moving across the sky during the nighttime.

But the moon is not moving across the sky.

The earth rotates and the moon follows a path.

In the evening the moon is in the east.

It seems to rise.

In the morning the moon is in the west.

It seems to set.

The Shape of the Moon

The moon moves around the earth.

The moon seems to change shape every month.

Sometimes you see only part of the moon.

You see the part that is reflecting the sun's light.

This part changes little by little each night.

The different shapes of the moon are called the **phases of the moon**.

What are the different shapes of the moon called?

Sometimes none of the moon is bright.

This is called a *new moon*.

Sometimes we see half the moon.

This is called a *quarter moon*.

Sometimes the moon looks like a big, bright ball.

This is called a *full moon*.

It takes about 29 days to see all the different phases of the moon.

New moon First quarter Third quarter Full moon

What does a full moon look like?

EXPLORATION

Process Skills
- Observe
- Predict

Lesson 53

Changes in the Sky

Our sky changes every day.
Sometimes we see the sun.
Sometimes we see other stars.
Sometimes we see the moon.
In this Exploration you will observe the nighttime sky and the daytime sky.
You will predict what the moon will look like in about one month.

Chapter 8

Big Question

What causes the seasons?

Seasons

God designed the earth to move.
The earth moves in different ways.
As it moves there are changes.
It may become colder or hotter.
There may be more daylight or less daylight.
The time of year may change, but God never changes.

The Earth Moves

It Rotates

A toy top spins.

A ball spins.

The earth spins.

Because the earth spins, or rotates, there is day and night.

We can see the sunrise and the sunset.

We can see the sun, moon, and stars.

Meet the Scientist

Nicolaus Copernicus was a scientist who lived long ago. He studied the heavens. He observed that the earth moves. He proved that the earth goes around the sun. It took many years for other scientists to believe this discovery.

It Revolves

The earth moves in another way.

The earth moves around the sun.

We say the earth revolves.

To **revolve** means to move around another object.

The earth takes one year to revolve around the sun.

A **year** is the time it takes the earth to make one trip around the sun.

How long does it take the earth to revolve around the sun?

The Earth Tilts

God designed the earth to tilt.

To **tilt** means to lean.

The earth always tilts in the same direction.

The earth is tilted as it rotates.

The earth is tilted as it revolves.

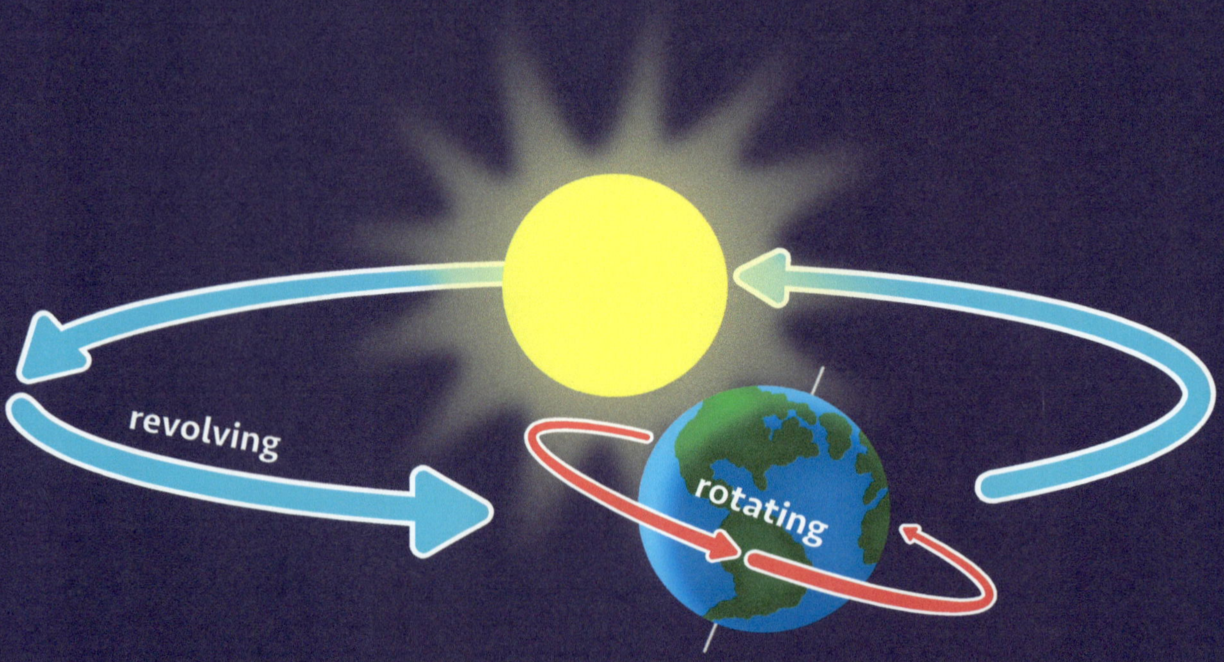

The earth's tilt causes seasons.
The earth's revolving around the sun also causes seasons.
A **season** is a time of year.
The seasons are *winter*, *spring*, *summer*, and *fall*.

Seasons

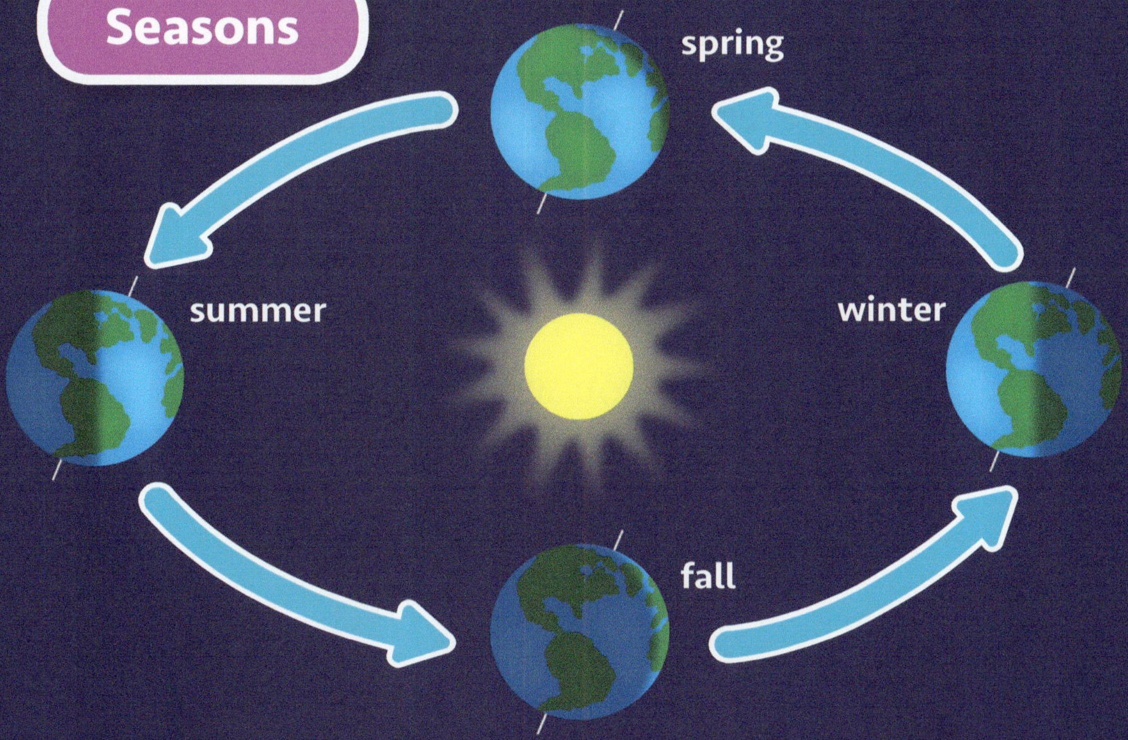

What two things cause seasons?

EXPLORATION

Process Skill

• Measure

Lesson 57

Using a Thermometer

How hot is it outside?

Temperature is how hot or cold something is.

A **thermometer** is a tool that measures temperature.

A thermometer tells how hot or cold something is.

In this Exploration, you will read and record temperature using a thermometer.

The Seasons

God gave us seasons.
In Genesis 8:22 God promises that there will always be seasons.
The seasons change.

God designed seasons to be a cycle.
Winter changes to spring.
Spring changes to summer.
Summer changes to fall.
And fall changes to winter again.

Winter

Winter comes after fall.
Winter begins in December.
The earth is tilted away from the sun.
The amount of daylight is shortest during winter.
This makes winter the coldest season.
Sometimes snow falls.

Winter is a time of resting.
Many trees have lost their leaves.
Some plants die.
Other plants rest.
They will start to grow again in spring.

God designed animals to survive during winter.

Some animals grow thicker fur to stay warm.

Some animals have fur that changes color in winter.

Some animals sleep through winter.

What is the difference between summer and winter fur?

Why is the temperature coldest in winter?

Spring

Spring comes after winter.
Spring begins in March.
The amount of daylight gets longer.
The temperature becomes warmer.
Spring rain comes.

Spring is a time of planting seeds.
The rain helps the seeds grow.
Flowers grow.
Trees grow new leaves.

Baby animals are born in spring.
Birds build nests and lay eggs.
Bees buzz around flowers.
The fur of some animals changes color in spring.

What happens to some trees in spring?

STEM Careers

Lesson 60

Landscape Architect

A landscape architect is a type of engineer.

A **landscape architect** helps people design their yards.

He can design parks and gardens.

He measures the space.

He chooses the right plants for that environment.

He can use rocks and fountains.

He draws a plan of his idea.

He shares his plan with others.

Summer

Summer comes after spring.
Summer begins in June.
The earth tilts toward the sun.
The amount of daylight is longest in summer.
This makes summer the hottest season.

Is daylight long or short in summer?

Summer is a time for growing.
Plants grow bigger.
Some trees and bushes grow fruit.
Vegetables grow and become ripe.

Baby animals eat and grow.
Baby birds begin to fly and leave their nests.

Fall

Fall comes after summer.
Fall begins in September.
The amount of daylight gets shorter.
It gets dark earlier.
The temperature gets cooler.

Plants change in the fall.
Some leaves change colors.
Some leaves fall to the ground.
Some plants stop growing.
Fall is a time of harvest.
It is a time to pick fruits and vegetables.

Some animals gather food for winter.
Other animals move to a new place.
They go to a place that has more food.
They go to a place that is warmer.
The animals will return in spring.

How do animals prepare for winter?

EXPLORATION

Process Skills
- Infer
- Communicate

Lesson 61

Seasons Where I Live

The seasons are a cycle.
There may be four seasons where you live.
There may not be four seasons where you live.

What are seasons like where you live?
In this Exploration you will make a book.
Your book will show the seasons that you observe where you live.

Chapter 9

Big Question

How does the weather change from day to day?

Weather

Every day is different.
Will it be warm tomorrow, or will it be cold?
Will it rain today, or will it be dry?
If it is windy, you might fly a kite.
If it rains, you might splash in a puddle.
If it snows, you might build a snowman.
Living in God's world is exciting.

Lesson 64

The Weather

It is hot.

It is windy.

It is snowing.

These are ways you think about the weather.

The **weather** is what the air outside is like.

The weather is a part of God's world.
The weather changes from day to day.
One day may be sunny and clear.
The next day may be cloudy and rainy.

What is weather?

Temperature

Temperature is how hot or cold it is.

A thermometer is used to measure the temperature of the air.

The thermometer has numbers on it.

The numbers show the temperature of the air in degrees Fahrenheit or Celsius.

The temperature changes during the day.

In the daytime the sun warms the air.

The temperature goes up.

At nighttime the sun does not warm the air.

The temperature goes down.

How does temperature change during a day?

Wind

Wind is moving air.
Warm air rises.
Cool air sinks.
The differences in the air cause wind.

Wind moves things.
It moves leaves on trees.
It moves sailboats.

Wind also moves flags.
You can look at a flag to see how hard the wind is blowing.

You may say the wind is *calm*.
The wind does not blow much.

You may say there is a *light* wind.
The wind blows a little.

You may say the wind is *strong*.
The wind blows a lot.

What is wind?

The Water Cycle

The water moves from the earth to the sky and back to the earth again. It moves this way over and over. This is called the **water cycle**.

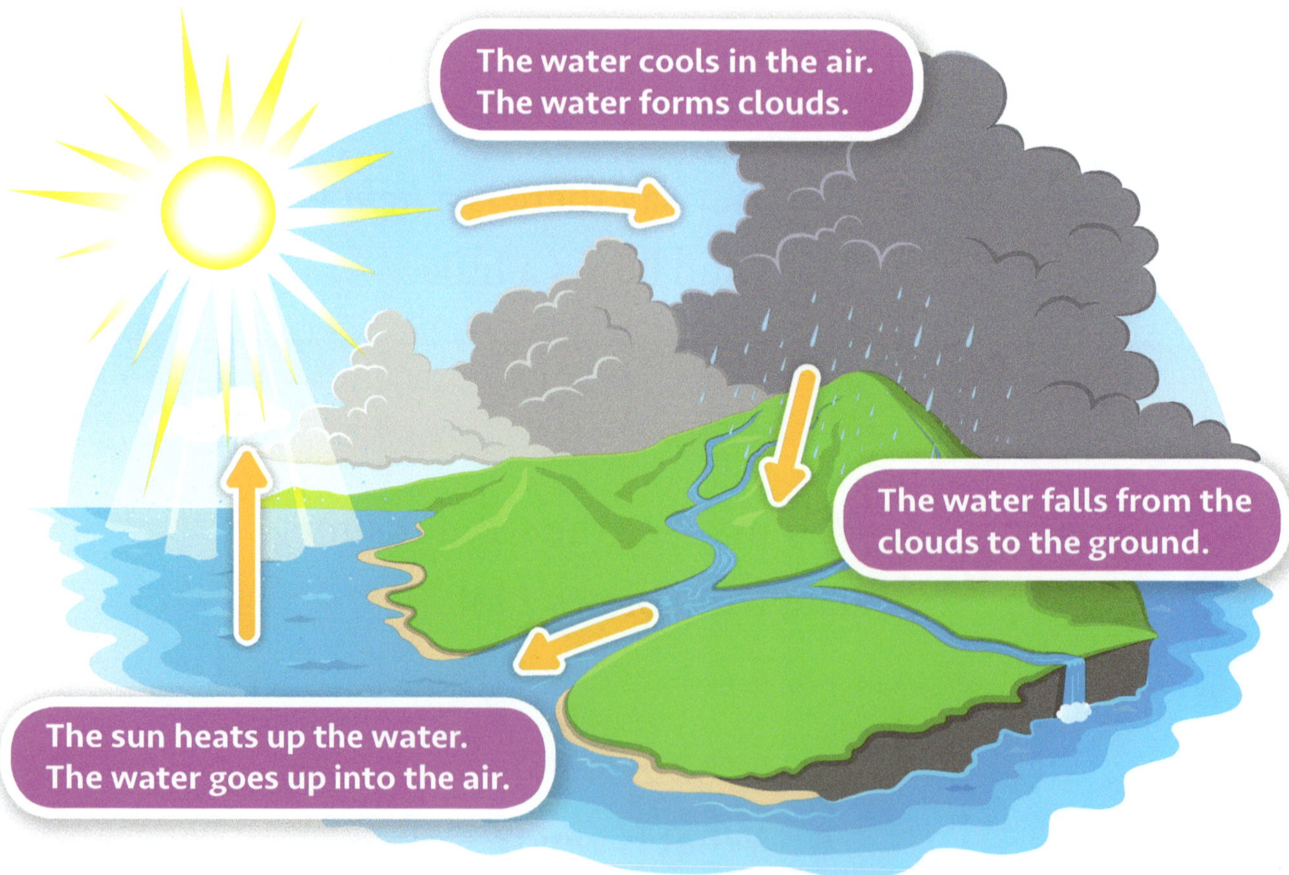

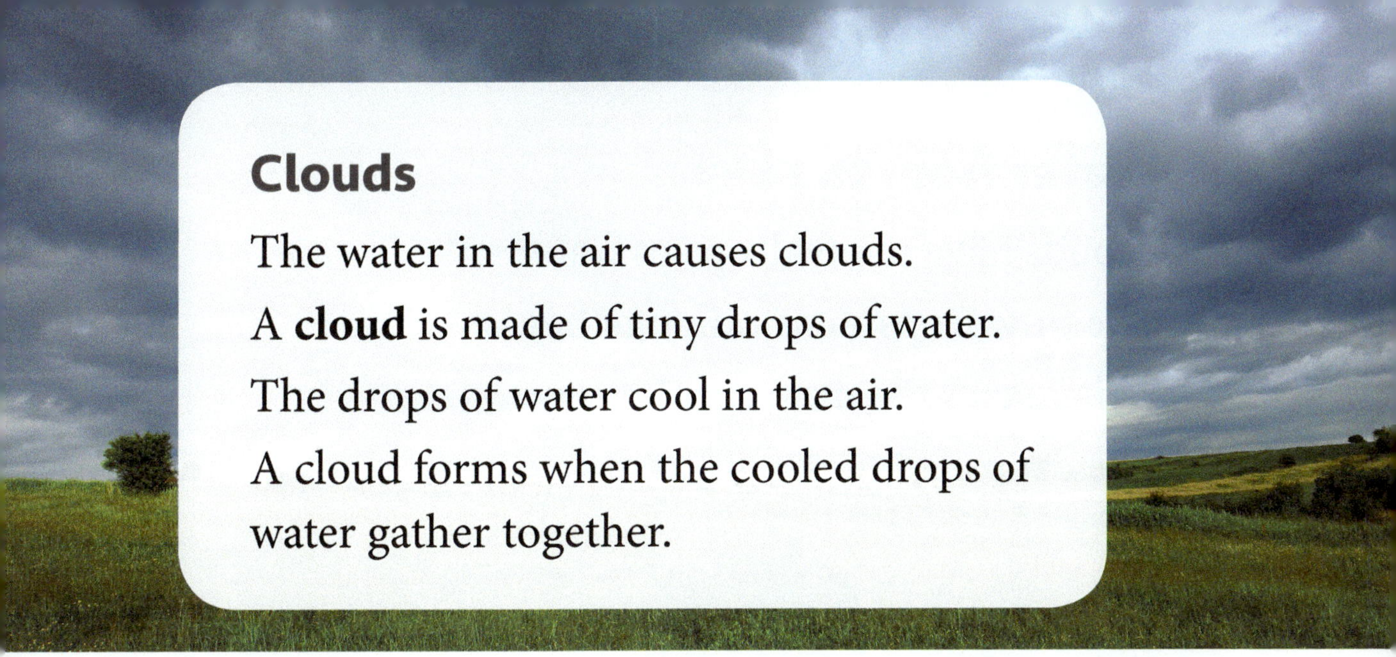

Clouds

The water in the air causes clouds.

A **cloud** is made of tiny drops of water.

The drops of water cool in the air.

A cloud forms when the cooled drops of water gather together.

You observe clouds in the sky.

You may say it is *clear*.

There are no clouds.

You may say it is *partly cloudy*.

There are some clouds.

You may say it is *cloudy*.

There are many clouds.

What causes clouds to form?

Water from the Sky

Clouds have many drops of water.
The tiny drops may join together.
This makes the drops get heavy.
The drops fall to the ground.

You may say the water that falls from the sky is *rain*.

You may say it is *sleet*.

You may say the water that falls from the sky is *hail*.

You may say it is *snow*.

What are four kinds of water that fall from the sky?

STEM Careers

Lesson 66

Meteorologist

A **meteorologist** is a scientist who studies the weather.

He uses his senses to find out about the air outside.

A meteorologist *observes* the weather.

He uses tools to observe the weather.

A meteorologist *predicts* the weather after he observes it.

He makes a careful guess at what may happen.

A meteorologist predicts the weather to help other people.

A meteorologist *records* what he observes about the weather. He records, or writes down, what he finds out. He makes maps and charts.

A meteorologist then *reports* the weather. He reports, or tells, what he predicts. A meteorologist tries to be trustworthy. Many times what he predicts happens. Sometimes what he predicts does not happen.

What four things does a meteorologist do?

EXPLORATION

Lessons 67–68

Weather Watching

Process Skills
- Observe
- Predict

The weather changes every day.
God tells us to rule over the earth.
A meteorologist studies the weather.
When he predicts the weather, he helps others to be safe.
You can be a meteorologist too.
In this Exploration you will observe and predict the weather.

Science and the Bible

During the Genesis Flood, it rained for many days. Water covered the whole earth. After the Flood, God made a promise. He said that He would never again cover all the earth with water. He put a rainbow in the sky. Noah and his family saw the rainbow. The rainbow reminds us that God keeps His promises.

Let’s Learn About

5

Energy

Chapter 10

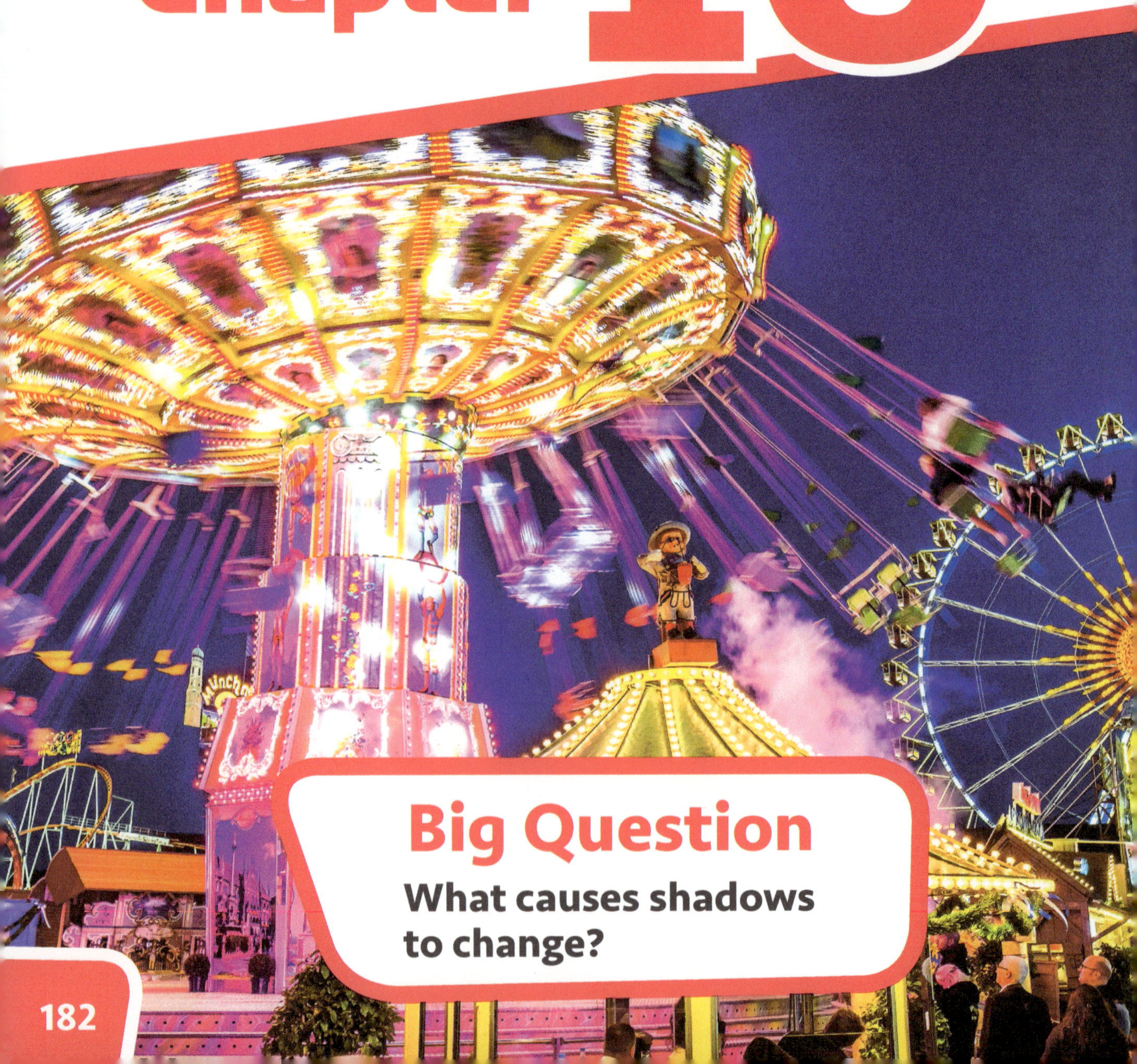

Big Question

What causes shadows to change?

Light

What is the first thing you do when you enter a dark room?

What helps you see to play a sport at night?

What do you see high in the sky on a bright day?

People, animals, and plants all need light to survive and grow.

God tells people to care for the earth.

Studying light helps you use light well.

Using light well helps you care for the earth.

What Light Is

Food gives plants and animals energy.
Food gives you energy.
Energy is what is needed to cause change.

Light is a form of energy that people need in order to see.
The Bible says that God created the sun for the daytime.
God created the moon and stars to rule the night.

Where Light Comes From

Where does light come from?

Where will you look for that answer?

What Some Scientists Say

You look in a science book in the library.

You read that many, many, many years ago there was a tiny speck.

The speck went *bang!*

Then the speck got bigger and bigger.

There was light energy in the speck.

Some scientists explain that light came from this tiny speck.

What is light?

What the Bible Says

You look for the answer in the Bible.

You read Genesis 1.

It says, "And God said, Let there be light: and there was light."

Light came from God speaking.

When science does not agree with the Bible, the Bible is right.

The Bible is true because it is God's Word.

Natural or Manmade

Some light is natural.
Natural lights are made by God.
The sun and the stars are natural lights.
The sun and the stars make their own light.

God gave people the minds to make light.
This kind of light is called manmade.
Flashlights, lamps, and candles are manmade.
These objects do not make their own light.
Man has made the light in each of them.

Is your bedroom light a natural or manmade light?

STEM Careers

Electrician

Many manmade lights use electricity.
Electricity is a form of energy.
Because of electricity you can read a book at night.
A lamp uses electricity to give you light.
Because of electricity you can watch TV.

An **electrician** works with electricity.
He designs electrical systems for homes, stores, and offices.
He puts in the electrical wires.
An electrician keeps them working.
He knows how to work safely.
An electrician's job is very important.

INVESTIGATION

Process Skills
- Predict
- Observe
- Classify

Lesson 72

Observing Light

Light travels through some objects.
Light travels through a window.
Some objects block light.
Light does not travel through your desk.

In this Investigation, you will shine a flashlight on some objects.
You will observe how the light travels through the objects.
You will classify the objects by how much light travels through them.

Light and Objects

Transparent

Light can travel through some objects. Anything that allows light to travel through it is **transparent**.

Air and water are transparent. Light travels through them. Clear plastic and clear glass windows are transparent. Light travels through them.

Translucent

Anything that allows some light to travel through it is **translucent**.

Tissue paper is translucent.

Sunglasses and wax paper are translucent.

Some light travels through these objects.

Some light is blocked.

How much light travels through a translucent object?

Opaque

Some objects block all light.

Anything that does not allow light to travel through it is **opaque**.

Light does not travel through this book.

Light does not travel through you.

All light is blocked.

Your book and your body are both opaque.

Shadows

A **shadow** is a dark shape formed when an object blocks light.
The place where the shadow forms depends on where the light is.

Shine a light on the right side of an object.
The shadow forms on the left side.

Shine a light on the left side of an object.
The shadow forms on the right side.

What kind of object blocks all light?

A shadow has a shape like the object that blocks the light.

The sun is in different places during the day.

You can see shadows change.

In the morning the sun is low in the eastern sky.

This makes long shadows.

In the middle of the day, the sun is high above. The shadows are short.

By evening, the sun is low in the western sky. The shadows are long again. Now the shadows are in a different place than they were in the morning.

During the day, when is your shadow the shortest?

INVESTIGATION

Lesson 74

Illuminate Objects

Process Skills
- Predict
- Observe
- Infer

When you turn a light on in a room, it **illuminates**, or lights up, the room. The sun illuminates our daytime. A table lamp illuminates a desk.

In this Investigation, you will predict whether an object is illuminated by its own light or by another light. You will use a pinhole box to observe different objects.

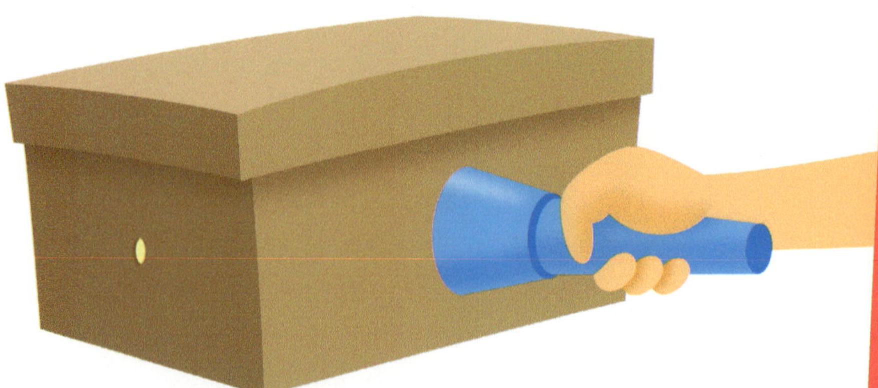

Light Illuminates

Some objects are illuminated by their own light.

Most objects do not make their own light.

You see these objects because they are illuminated by another light.

How Light Travels

Light travels in a straight line.

Light hits an object.

Then the object reflects the light.

Your eyes see the light that reflects off the object.

Health and Safety

God made your eyes. You should take care of them. Sunlight travels to your eyes. It can reflect off shiny things. Looking at the sun can damage your eyes. You should wear a hat and sunglasses. Both of these will keep most of the sunlight from your eyes.

Shiny and smooth objects reflect light well.
The shiny parts of cars reflect sunlight.
The rain on a road reflects light.
The moon reflects the light of the sun.

A **mirror** is shiny glass that reflects light.
Light travels from an object to the mirror.
The mirror reflects the image back to your eyes.

Light is a gift from God that allows us to observe His world.
Learning about light will help us care for the earth.

How does light travel?

Chapter 11

Big Question

How does sound change?

Sound

Boom!

Drip.

Splash!

Ring.

You hear noises all around you.

What are you hearing?

How do you hear them?

We will find out!

Lesson 78

How Sound Is Made

You hear sounds with your ears all the time.

A dog barks.

A phone rings.

A friend laughs.

People cheer.

Hearing sounds helps you observe God's world.

Sound is a form of energy that you can hear.
Sound is made when something vibrates.
To **vibrate** means to move back and forth very quickly.
Sound is a vibration that you can hear.

How can you make something vibrate?
You can hit it or rub it.
You can shake it or pluck it.
You can blow air through it.

What is sound?

How Sound Moves

Sound Waves

When you make a sound, you can hear it.

You cannot see sound moving.

But you can hear it moving.

Sound moves, or travels, in waves.

Sound waves travel up and down.

Sound waves travel to the right and the left.

Sound waves travel in front of you and behind you.

Sound waves travel in all directions at the same time.

In what directions do sound waves travel?

Sound and Matter

Sound travels through matter.

Matter is anything that has weight and takes up space.

Matter is all around you.

Your toys are matter.

Your food is matter.

Your body is matter.

Even the air you breathe is matter.

Most of what we hear is sound that moves through the air.

Sound and Your Ears

Proverbs 20:12 tells us that God made your ears.

He designed your ears to hear sound.

Vibrating, or moving, air makes sound.

The moving air makes the inside of your ear vibrate.

Your brain knows what the vibrating inside your ear means.

Your brain tells you what you heard.

You are "fearfully and wonderfully made" (Psalm 139:14).

What did God design your ears to do?

How Sound Changes

Volume

Sound can change.

Volume is how loud or soft a sound is.

Loud sounds are easy to hear.

A shout is a loud sound.

Some loud sounds may hurt your ears.

Soft sounds may be hard to hear.
A whisper is a soft sound.
Soft sounds will not hurt your ears.

What is volume?

Health and Safety

God made your ears. Your ears are important to God. You should take care of your ears. You should keep them clean. You should protect them from loud sounds. You should protect them from the cold and wind. Caring for your ears pleases God.

Pitch

Pitch is how high or low a sound is.
Every sound has pitch.

The meow of a kitten is a high pitch.

The roar of a lion is a low pitch.

A musical instrument can make sounds with high pitches and low pitches.

You can sing in a high or a low pitch.

What is pitch?

INVESTIGATION

Lesson 81

Hearing Pitch

Process Skills

- Observe
- Measure
- Infer

Shhhh.

Listen.

There are sounds all around you.

Some sounds are loud.

Other sounds are soft.

Sounds are also high or low.

In this Investigation, you will find out what causes a rubber band to have a high or a low pitch.

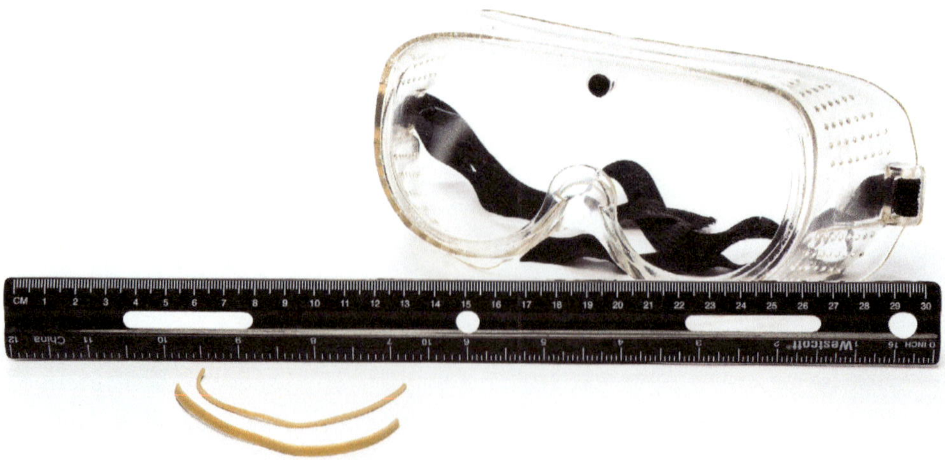

STEM

Lesson 82

Making Music

Some sounds make you happy.
Music can be a happy sound.

You may hum to make music.
Sometimes you may sing.
You may also play an instrument.

Music has high and low pitches.
In this STEM activity, you will design a stringed musical instrument.
You will design each string to have a different pitch.

Chapter 12

Big Question

How do we use light and sound to communicate?

Communicating with Light and Sound

You hear sirens.

You see flashing lights.

A fire truck drives down your street.

You may think there is a fire nearby.

Different lights and sounds let you know something is happening.

Lights and sounds can help people make safe choices.

Lesson 85

Around My Home

Light and sound are all around you.

Many objects use light and sound to communicate.

To **communicate** means to give information about something.

Light

A light on the stove lets you know it is on.

The light lets you know the stove may be hot.

People turn on their porch light at night.

The light helps them to see who is at their door.

Sound

Ding dong!

A doorbell makes this sound.

It lets you know someone is at your door.

An alarm clock beeps.

The sound lets you know it is time to wake up.

Some sounds are important for you to listen to.

Your parents tell you how to behave.

You should hear and obey what they say.

The Bible tells children that it is right to listen to their parents.

What sounds at home are important for you to listen to?

Light and Sound

A TV uses light and sound.
It communicates the words and pictures of your favorite show.

A cell phone uses light and sound.
The sound lets you know someone is calling.
The light lets you see who is calling.

Around My School

There are many objects in your school that communicate with light and sound.

Light

An exit sign uses light. This sign helps you find your way out of a building.

Light on a screen helps you see pictures. These pictures help you learn more about God's world.

Which kind of sign helps you find your way out of a building?

Sound

You may hear a bell ring in your classroom.
This sound lets you know that it is time to begin the day.

Your teacher uses her voice to teach you about God's world.
Your teacher reads a book to you.
She blows her whistle at recess.
You know it is time to line up!

Light and Sound

A computer uses light and sound to communicate.

The light lets you see a video.

The sound lets you listen to the words.

Your classroom has a fire alarm.

It uses light and sound to let you know there may be a fire.

The light and sound mean that you need to leave the building.

Your class presents a special program.

Light and sound help you communicate a message to others.

How do light and sound help you at school?

Around My Community

Light and sound help you make safe choices in your community.

God wants people to use light and sound to help others.

God gave people minds to understand what light and sound communicate.

Light

A traffic light lets drivers know when to stop and go.

A railroad crossing uses flashing lights to let people know a train is coming.

Sound

Some sounds do not use words but still tell you something.

A tornado siren warns people of danger.

It tells people that bad weather may be coming.

A hissing snake or a growling dog tells you to stay away.

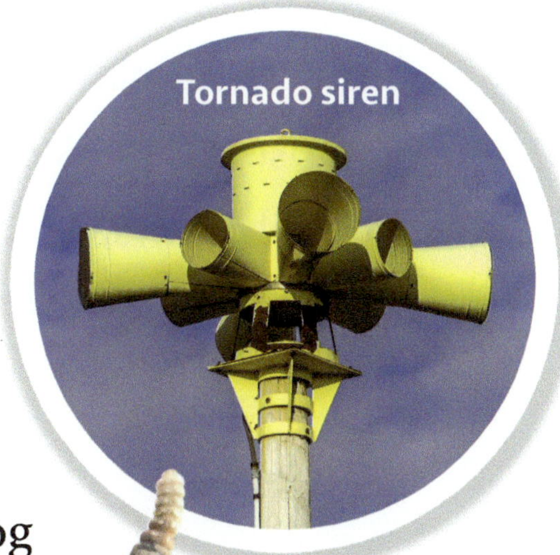

Tornado siren

You can enjoy different sounds in your community.

A band plays music as it marches down your street.

People cheer as they watch your soccer game.

How do light and sound in your community keep you safe?

You can also use sound to praise God.
Your friend plays the piano to make music.
You use your voice to sing at church.
The Bible says, "It is good to sing praises unto our God" (Psalm 147:1).

Light and Sound

A police car uses flashing lights and sounds.

It tells a driver to stop.

An ambulance uses flashing lights and sounds.

It tells drivers to move out of the way.

Light and sound can also help you celebrate.

You can see and hear fireworks on the Fourth of July.

Fireworks are bright and loud.

Why does a police car use lights and sounds?

Fantastic Facts

Lighthouses are built near the ocean. They use light and sound to communicate with ships at sea. Many ships travel at night. The lighthouse shines a bright light. The light helps guide captains. Heavy fog can make it hard to see the light. A lighthouse also uses a fog horn. It is a loud alarm. This alarm warns the captain of danger he cannot see. A lighthouse uses light and sound to help ships travel safely in the ocean.

STEM

Lesson 87

Helping with Light or Sound

Your friend has a need.

You see a light.

You hear a sound.

You have an idea!

In this STEM activity you will design a way to use light or sound to help someone in need.

Biblical Worldview

Everyone has a worldview.

You have a worldview.

You have studied science to learn about God's world.

A worldview helps explain what you learn.

It is important to study science with a biblical worldview.

Why do you study science?

Where did our world come from?

God created a perfect world in six days.

He created the plants and animals.

He created the sun, moon, and stars.

God created light and sound.

God created people in His image.

The Bible says we are to praise God.

We are to thank God for making the world.

We are to thank God for making us.

Why do things in our world work together the way they do?

God designed the parts of the earth to work together.

Some people see the weeds and thistles.

They see pain, sickness, and death.

These things happened because of Adam's sin, or the Fall.

What is the Fall?

The Bible teaches that God will make this world perfect again.

One day everything will work together as it should.

Jesus will rule the earth.

Everyone who turns away from sin and trusts in Jesus will live with God forever.

We call this **Redemption**.

Who are we? Why are we here?

Some books say that God did not make people.

These books teach that people came from nature.

They teach that nature is the mother of all living things.

God says that people are the most important part of His creation.

People are to love God more than they love anything or anyone else.

The Bible teaches that the greatest commandment is to love God with all our hearts.

We are to try to be like Jesus.

We are to love other people because they are made in God's image.

We are to take care of the earth that God has made.

What is the most important part of God's creation?

Why is science important?

Science can be a powerful tool to help us answer questions.

Some books teach that science can answer all important questions.

They say that science is the only way to know something for sure.

Some books teach that science proves the Bible is wrong.

They even say that science proves there is no God.

Science is good at answering some questions.

What are the needs of plants?

Why does the sun seem to travel across the sky?

Why are there seasons?

How are light and sound used to communicate better?

What kinds of questions can science answer?

There are other important questions that science cannot answer.

Who is God?

Why am I here?

Where did light come from?

Did a flood cover all the earth with water?

The Bible can answer these questions.

What the Bible says is true.

When science disagrees with the Bible, the Bible is right.

What should you believe when science disagrees with the Bible?

EXPLORATION

Process Skills
- Observe
- Communicate

Lesson 90

A Song of Praise

You can sing songs to praise God.

You can play music to praise Him.

You praise God to thank Him.

You thank God for who He is and what He does.

In this Exploration you will write a song.

Your song will praise God for His creation.

Science and the Bible

Music is one way people communicate with sound. There are many songs in the Bible. The book of Psalms is filled with songs of praise. King David wrote many of the psalms. Sometimes music was played when they were sung. It is good to praise God with our music.

Glossary

A

adult plant—A fully grown plant that can make more plants.

air—A need of living things.

arm—The body part that reaches for things.

astronaut—A person who travels into space.

B

balance scale—A tool used to measure the weight of something or how heavy something is.

bird—An animal that has feathers and wings.

bones—The hard parts of the body.

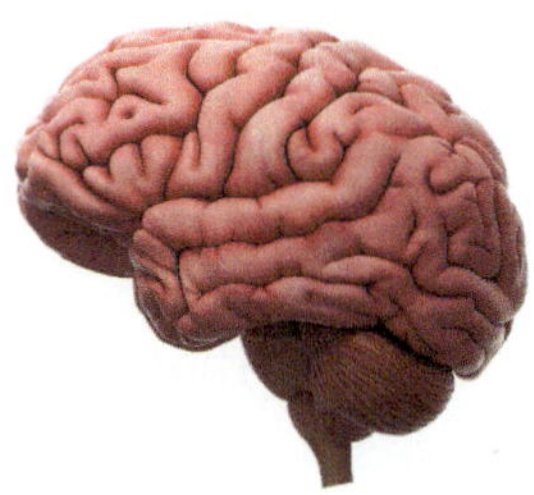

brain—The body part that controls everything in the body.

C

calm wind—Wind that does not blow much.

cavity—A hole in a tooth.

classify—To group things that are alike.

clear—No clouds in the sky.

cloud—Many tiny drops of water that have cooled in the air and gathered together.

cloudy—Many clouds in the sky.

communicate—To share what you know; to give information about something.

Creation—The belief that God spoke and made all things.

D

day—**1.** The time that the earth takes to rotate one time. **2.** What God called the light at Creation. **3.** The time between sunrise and sunset.

dentist—A doctor who takes care of teeth; a STEM career.

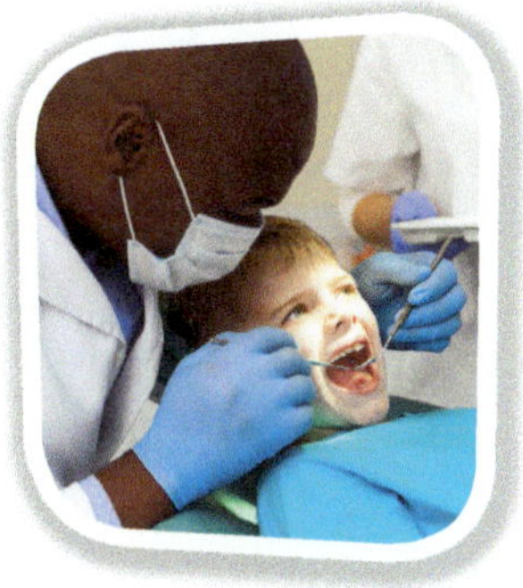

design—To plan something.

dirt—A need of plants.

E

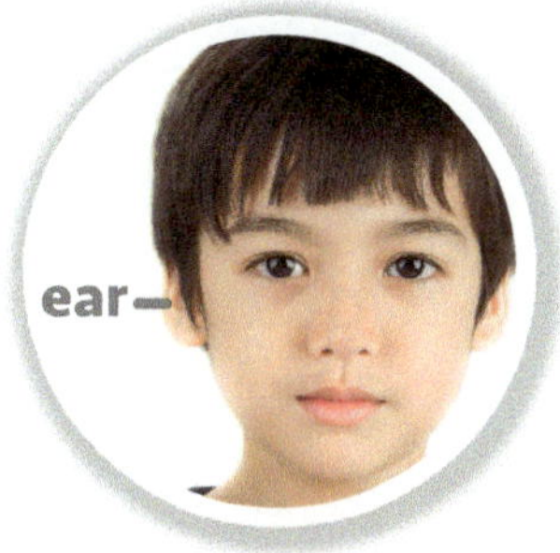

ear—The body part on the head that allows you to hear.

electrician—A person who works with electricity; a STEM career.

electricity—A form of energy.

energy—What is needed to cause change.

engineer—A person who uses math and science to answer everyday problems; a STEM career.

environment—All the living and nonliving things around a plant or animal.

eye—The body part on the head that allows you to see God's creation.

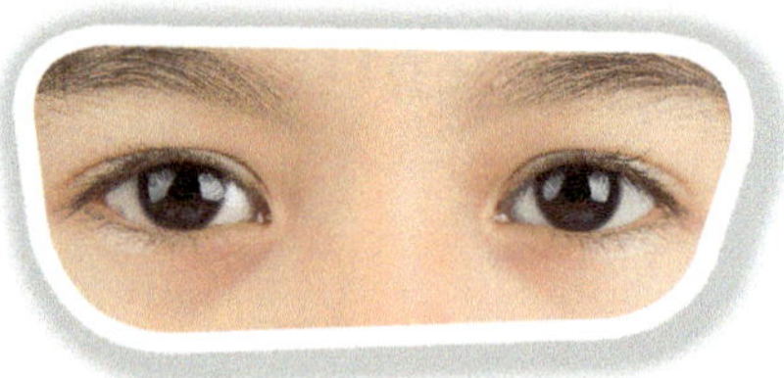

F

Fall—The event described in Genesis 3 when the first man, Adam, sinned and brought death into the world.

fall—The time of year that comes after summer.

fish—An animal that lives in water and has gills.

flower—The part of the plant that makes seeds.

food—A need of living things.

full moon—Moon phase when the moon looks like a big, bright ball.

G

germs—Things that can make a person sick.

H

habit—Something a person does over and over again.

hail—A type of water that falls from the sky.

hand lens—A science tool that makes small things look bigger.

head—The body part that holds the brain.

healthy—Not sick or hurt.

hearing—The sense that lets you hear with your ears.

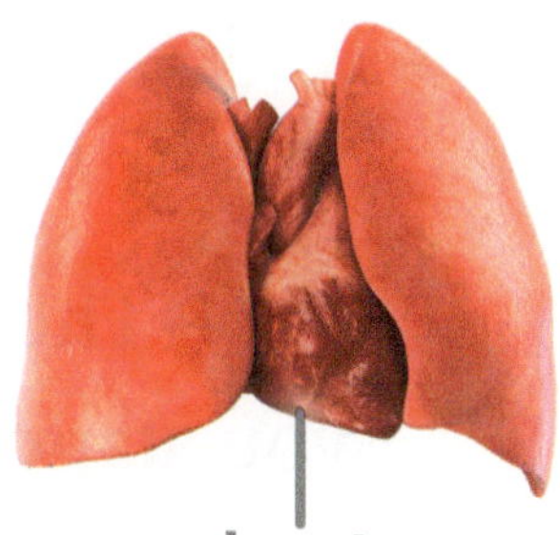

heart—The body part that pumps blood through the body.

hypothesis—A possible answer to a question.

I

illuminate—Light up.

image—Likeness.

infer—To use what you know to tell why things happen.

investigation—A scientific test to find an answer to a question.

L

landscape architect—An engineer who designs spaces for plants; a STEM career.

leaf—The part of a plant that makes food for the plant.

leg—The body part that helps you stand, walk, and run.

life cycle—All the steps of the life of a living thing.

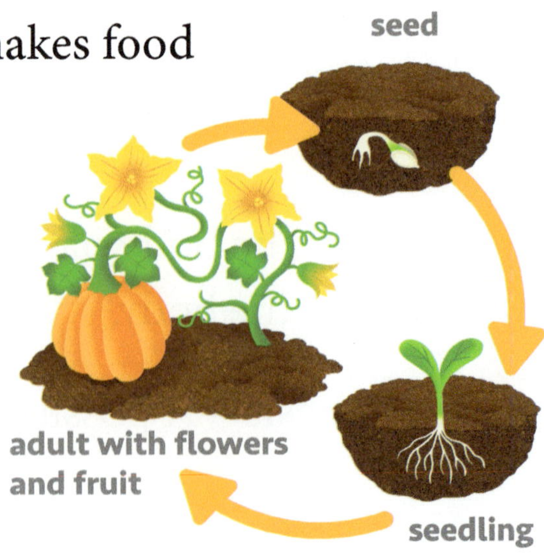

light—**1.** What a plant needs to make food. **2.** A form of energy that people need in order to see.

light wind—Wind that blows a little.

living—Something that needs food, water, air, and space to live and grow.

lung—The body part that fills with air when you breathe in.

M

mammal—An animal that has fur or hair.

matter—Anything that has weight and takes up space.

measure—To find the size, number, or amount of something.

measuring cup—A tool that measures how much space something takes up.

meteorologist—A scientist who studies the weather; a STEM career.

mirror—A shiny glass that reflects light.

moon—A big ball of rock that moves around the earth.

mouth—The body part on the head that allows you to speak, sing, and eat.

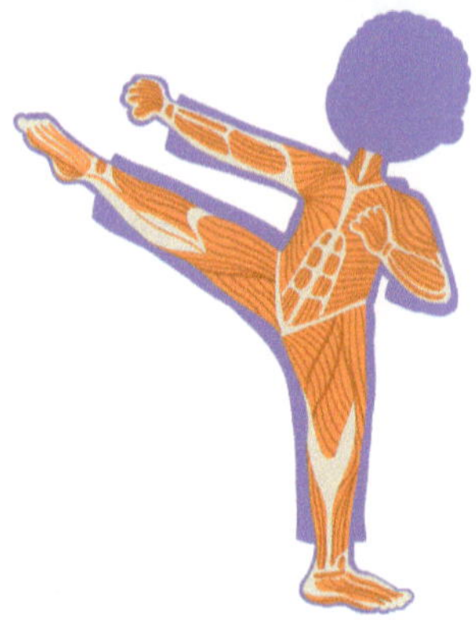

muscles—Body parts that make the bones and other body parts move.

N

need—What a living thing must have to stay alive and grow.

new moon—Moon phase when none of the moon is bright.

night—**1.** What God called the darkness at Creation. **2.** The time between sunset and sunrise.

nonliving—Something that does not need food, water, and air.

nose—The body part on the head that allows you to smell.

O

observe—To use your senses to find out something.

offspring—Babies.

opaque—Anything that does not allow light to travel through it.

P

partly cloudy—Some clouds in the sky.

phases of the moon—The different shapes of the moon.

pitch—How high or low a sound is.

praise—To thank God for who He is and what He does.

predict—To make a careful guess at what may happen after observing; using what you know to say what may happen.

Q

quarter moon—Moon phase when only half the moon is bright and can be seen.

R

Redemption—God's work of saving from sin and making His creation perfect again.

reflect—To bounce off.

report—To tell others about something.

revolve—To move around another object.

root—The part of a plant that holds the plant in the ground and takes in water.

rotate—To spin.

ruler—A tool that measures how long something is in inches or centimeters.

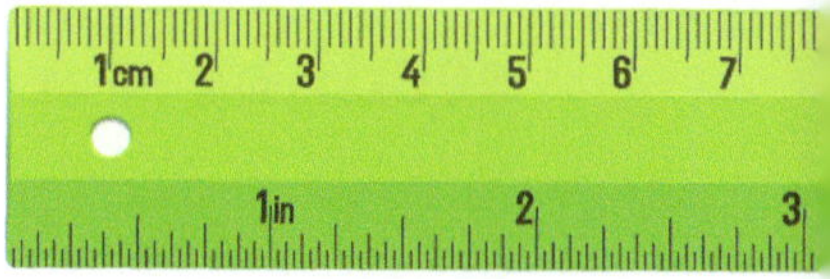

S

safe—Staying away from things that can hurt the body.

science—The study of God's world.

science process skill—A skill or action used to help a scientist discover and use facts.

science tool—An object that helps scientists work.

scientific method—A plan for doing an investigation.

scientist—A person who uses his senses to study God's world.

season—A time of year.

seed—The beginning of a plant's life cycle; what many plants grow from.

seedling—A young plant.

senses—The ways God gave you to find out about the world.

shadow—A dark shape formed when an object blocks light.

shelter—A safe place to live; a need of living things.

sight—The sense that lets you see with your eyes.

sleet—A type of water that falls from the sky.

smell—The sense that lets you identify scents with your nose.

snow—A type of water that falls from the sky.

sound—A form of energy that you can hear.

space—A need of living things.

spring—The time of year that comes after winter.

star—An object in the sky that makes its own light.

stem—The part of a plant that holds up the plant and moves water from the roots to the other parts of the plant.

stomach—The body part that breaks down the swallowed food.

strong wind—Wind that blows a lot.

summer—The time of year that comes after spring.

sun—The closest star to the earth; the star that warms and lights the earth.

sunrise—The time when the sun can first be seen in the morning.

sunset—The time when the sun begins to disappear in the evening.

survive—To stay alive.

T

taste—The sense that lets you identify flavors with your tongue.

telescope—A tool that makes far away things look closer and bigger.

temperature—How hot or cold something is.

thermometer—A tool used to measure temperature.

tilt—To lean.

touch—The sense that lets you feel with your hands and fingers.

translucent—Anything that allows some light to travel through it.

transparent—Anything that allows light to travel through it.

V

vibrate—To move back and forth very quickly.

volume—How loud or soft a sound is.

W

water—A need of living things.

water cycle—Water that moves from the earth to the sky and back to the earth again.

weather—What the air outside is like.

wind—Moving air.

winter—The time of year that comes after fall.

worldview—The way a person thinks about the world.

Y

year—The time it takes the earth to make one trip around the sun.

Z

zoologist—A scientist who studies and takes care of animals; a STEM career.

Index

A

B

C

D

E

F

G

H

N

O

P

Q

R

S

T

V

W

Y

Z

Photo Credits

Key: (t) top; (c) center;
(b) bottom; (l) left; (r) right;
(bg) background; (i) inset

Cover

front Toucan © iStock.com/edurivero; Rusty gears © iStock.com/GLYPHstock; Caterpillar © iStock.com/Captainflash; **back** © iStock .com/Mustafa Kocabasi

Unit Openers

1l Butterfly Hunter/Shutterstock.com; **1** (wing) Sukpaiboonwat/Shutterstock.com; **1** (magnifying glass) © iStock.com/kyoshino; **44–45** Milan Zygmunt/Shutterstock.com; **88–89** Puwadol Jaturawutthichai/Shutterstock .com; **124–125** Lambros Kazan/Shutterstock .com; **180–181** Thinkstock/Stockbyte /Thinkstock

Chapter 1

2 Janie Airey/Digital Vision/Getty Images; **4** Cathy Yeulet /123RF; **5** PhotoDisc, Inc.; **6** ALPA PROD/Shutterstock.com; **7** Monkey Business Images/Shutterstock.com; **8**t Jiri Hera/Shutterstock.com; **8**b photosbysuzi /Bigstock.com; **9** Mila May/Shutterstock.com; **10**t PhotoDisc/Getty; **10**bl Seregam /Shutterstock.com; **10**br vitec/Shutterstock .com; **11**l Getty Images/Thinkstock; **11**r Getty Images/Hemera/Thinkstock; **13**t Darren Baker/Shutterstock.com; **13**c © iStock.com /BartCo; **13**b © iStock.com/bluecinema; **14** © iStock.com/StefaNikolic; **15**t © iStock.com /nicolas_; **15**b Blue Jeans Images/Media Bakery **17** "Spinophorosaurus holotype" by Remes K, Ortega F, Fierro I, Joger U, Kosma R, et al. (2009) A New Basal Sauropod Dinosaur from the Middle Jurassic of Niger and the Early Evolution of Sauropoda. PLoS ONE 4(9): e6924.doi:10.1371/journal.pone.0006924 /Wikimedia Commons/CC By 2.5; **19** © iStock.com/wragg; **20** © iStock.com/vgajic; **21** © iStock.com/picture

Chapter 2

22–23 Chutima Chaochaiya/Shutterstock.com; **24**l © iStock.com/3sbworld; **24**c grafvision /Shutterstock.com; **24**r–**25**l © iStock.com /emholk; **25**c © iStock.com/AfricaImages; **25**r Asia Images Group/Shutterstock.com; **26**, **27**b, **29**, **35**, **39**, **40** both, **42** BJU Press; **27**t © iStock.com/Ivantsov; **27**ct, **33**t, **37**t © piai - Fotolia.com; **27**cb Fedor Bobkov /Shutterstock.com; **28** © iStock.com /ayala_studio; **30** © iStock.com/FatCamera; **31** Elnur/Shutterstock.com; **32** Violart /Shutterstock.com; **33**t © piai - Fotolia .com; **33**b © iStock.com/RichVintage; **34**, **37**bc Raland/Shutterstock.com; **36**, **37**bl VladisChern/Shutterstock.com; **37**br © iStock .com/kyoshino

Chapter 3

46–47 Walter Quirtmair/Moment/Getty Images; **50** Cristian M Balate/Shutterstock .com; **51** taka1022/Shutterstock.com; **52**tl anne willette/Shutterstock.com; **52**tr Hurst Photo/Shutterstock; **52**bl Moko Photography /Shutterstock.com; **52**br © 2009 JupiterImages Corporation; **53**tr Tamara Kulikova /Shutterstock.com; **53**i © iStock.com/Kazakov; **53** © iStock.com/Difydave; **54** Anton Foltin /Shutterstock.com; **55** BJU Photo Services; **56** Zaitsava Olga/Shutterstock.com; **57**t Monkey Business Images/Shutterstock.com; **57**b © iStock.com/wingmar; **58**tl domnitsky /Shutterstock.com; **58**tr © iStock.com /joakimbkk; **58**c Noah Strycker/Shutterstock .com; **58**bl Blue Jeans Images/Media Bakery; **58**br © iStock.com/undefined undefined; **59**t harmpeti/Shutterstock.com; **59**b © iStockphoto.com/IgorDutina; **60** BJU Press; **61**l Imageman/Shutterstock.com;

61r Denis Pogostin/Shutterstock.com; **62**bg golubovystock/Shutterstock.com; **62**i mikeledray/Shutterstock.com; **64**t images72 /Shutterstock.com; **64**b "George Washington Carver c1910"/Wikimedia Commons/Public Domain

Chapter 4

66–67 Andrea Izzotti/Shutterstock.com; **70** © iStock.com/craftvision; **71**t © iStock.com /avs_lt; **71**b Schippers/Shutterstock.com; **73**t © iStock.com/ImagineGolf; **73**bl Cheryl E. Davis/Shutterstock.com; **73**br © iStock.com /paulafrench; **74**l © iStock.com/sturti; **74**r Eric Isselee/Shutterstock.com; **75**t ventdusud /Shutterstock.com; **75**b PAUL ELLIS/ Staff/AFP via Getty Images; **76** Damsea/ Shutterstock.com; **77**l Brian E Kushner/ Shutterstock.com; **77**r Rich Carey/ Shutterstock.com; **78**t © iStock.com/ Abramova_Kseniya; **78**b © iStock.com/Zoran Kolundzija; **79**t © iStock.com/photomaru; **79**b kajornyot wildlife photography/Shutterstock. com; **80** © iStock.com/PaulMoyer; **81**t JPBC/ Shutterstock.com; **81**b cpaulfell/Shutterstock .com; **82**l © iStock.com/Andrey Danilovich; **82**c © iStock.com/anankkml; **82**r © iStock .com/ACS15; **83**l cynoclub/Shutterstock.com; **83**r MINT IMAGES/FRANS LANTING/ Media Bakery; **84**t © iStock.com/SolStock; **84**b © iStock.com/Calinat; **85** Brenda Carson/Shutterstock.com; **86**t Nataly Studio/Shutterstock.com; **86**b Jan Havlicek/ Shutterstock.com; **87**t © iStock.com/eve_ eve01genesis; **87**c (acacia leaves) xpixel /Shutterstock.com; **87**c (acacia leaves with flowers) Scisetti Alfio/Shutterstock.com; **87**bl Eric Isselee/Shutterstock.com; **87**br Tamara Kulikova/Shutterstock.com

Chapter 5

90–91 © iStock.com/ArtMarie; **92–93**t © iStock.com/GeorgeManga; **92**b Monkey Business Images/Shutterstock.com; **94** © iStock.com/BassittART; **95**t Samuel Borges Photography/Shutterstock.com; **95**b prometeus/Bigstock.com; **96** Littlekidmoment /Shutterstock.com; **97**t © iStock.com/kgfoto; **97**b SpeedKingz/Shutterstock.com; **98**tl Maria Sbytova/Shutterstock.com; **98**tr Phillip Dyhr Hobbs/Shutterstock.com; **98**c DenisNata /Shutterstock.com; **98**bl Nagy-Bagoly Arpad /Shutterstock.com; **98**br Cherry-Merry /Shutterstock.com; **99**t Africa Studio /Shutterstock.com; **99**b Secheltgirl/Shutterstock .com; **100** Fotos593/Shutterstock.com; **102**t Lisa S./Shutterstock.com; **102**b wavebreakmedia/Shutterstock.com; **104** u3d /Shutterstock.com; **105** Sebastian Kaulitzki /Shutterstock.com; **107**t greenland /Shutterstock.com; **107**c Rob Hainer /Shutterstock.com; **107**b © iStock.com/Tom Merton

Chapter 6

108–109 Sergey Novikov/Shutterstock.com; **112**t © iStock.com/monkeybusinessimages; **112**b Mat Hayward/Shutterstock.com; **113** © iStock.com/jorgeantonio; **114**l © iStock.com /bigmt; **114**r © iStock.com/timsa; **115**t tinna2727/123RF; **115**bl © iStock.com /nicolesy; **115**br Sergiy Bykhunenko /Shutterstock.com; **116** michaeljung /Shutterstock.com; **117** BJU Press; **121**l Joe Belanger/Shutterstock.com; **121**r Uppercut /Charlie Schuck/Media Bakery; **122** © iStock .com/og-vision; **123** ppart/Shutterstock.com

Chapter 7

126–127 PaulPaladin/Shutterstock.com; **131** NASA Goddard Space Flight Center Image by Reto Stöckli (land surface, shallow water, clouds). Enhancements by Robert Simmon (ocean color, compositing, 3D globes, animation). Data and technical support: MODIS Land Group; MODIS Science Data Support Team; MODIS Atmosphere Group; MODIS Ocean Group Additional data: USGS EROS Data Center (topography); USGS Terrestrial Remote Sensing Flagstaff Field Center (Antarctica); Defense Meteorological Satellite Program (city lights); **136** BJU Press;

137 © iStock.com/lucentius; **138** © iStock.com/TonyBaggett; **140** iStockphoto/Thinkstock; **141**t Lambros Kazan/Shutterstock.com; **141**b Apollo 14 Shepard/Nasa/Wikimedia Commons/Public Domain; **144** Indiloo/Shutterstock.com; **145** ESB Professional/Shutterstock.com; **145** (moon) Indiloo/Shutterstock.com

Chapter 8

146tl Solodov Aleksei/Shutterstock.com; **146–147**tr karinkamon/Shutterstock.com; **146**bl, **161**tl Anton Petrus/Shutterstock.com; **146**br 2happy/Shutterstock.com; **148** "Nikolaus Kopernikus"/Wikimedia Commons/Public Domain; **153** all Joel Blit/Shutterstock.com; **155**t Nature Picture Library/Alamy Stock Photo; **155**bl, br Jim Cumming/Shutterstock.com; **158**t © iStock.com/shotbydave; **158**b Scott E. Feuer/Shutterstock.com; **159**l © iStock.com/ideabug; **159**tr Ian 2010/Shutterstock.com; **159**br KMLynch/Shutterstock.com; **160**t © iStock.com/kali9; **160**b T.W. van Urk/Shutterstock.com; **161**tr Hurst Photo/Shutterstock.com; **161**c LilKar/Shutterstock.com; **161**bl WNstock/Shutterstock.com; **161**br Alfa Photostudio/Shutterstock.com; **162** © Bkushner | Dreamstime.com; **162**i BLFootage/Shutterstock.com

Chapter 9

164–165 Evgeny Atamanenko/Shutterstock.com; **166–167**bg Jurga Jot/Shutterstock.com; **166**i © iStock.com/Marilyn Nieves; **167**i kp33photography/Moment/Getty Images; **168**l © iStockphoto.com/quidnunc; **168**r Getty Images/Comstock Images/Thinkstock; **170** ohrim/Shutterstock.com; **171** Tetra Images/Media Bakery; **173** © iStock.com/bgfoto; **176**t Tim Ockenden/PA Images/Alamy Stock Photo; **176**b Joe Raedle/Staff/Getty Images News via Getty Images; **177**t Handout/Getty Images News via Getty Images; **177**b Boston Globe/Contributor via Getty Images; **179** © iStock.com/mulia

Chapter 10

182–183 © iStock.com/FooTToo; **188**t © iStock.com/Bouillante; **188**b Francescomoufotografo/Shutterstock.com; **189** BJU Photo Services; **190**t © iStockphoto.com/skodonnell; **190**b BIGANDT.COM/Shutterstock.com; **191**t © iStockphoto.com/Bike_Maverick; **191**bl © Alex Bramwell | Dreamstime.com; **191**br Brooke Becker/Shutterstock.com; **192**t © Vladimir Nenov/123RF; **192**c © iStock.com/cunfek; **192**b © iStock.com/Eerik; **193** both, **194**l © iStock.com/ra-photos; **194**tr Nino H. Photography/Moment/Getty Images; **194**br © iStock.com/simonkr; **198** © iStock.com/joakimbkk; **199**t © iStock.com/stevecoleimages; **199**b Eric Isselee/Shutterstock.com

Chapter 11

200–201 gary718/Shutterstock.com; **202**tr Andrey_Popov/Shutterstock.com; **202**cr, **210**l © iStock.com/GlobalP; **202**bl © iStock.com/gpointstudio; **202**br © iStock.com/Steve Debenport; **203**tl © iStock.com/Wavebreakmedia; **203**tr © iStock.com/track5; **203**c © iStock.com/princessdlaf; **203**bl © iStock.com/jeecis; **203**br Martin Novak/Shutterstock.com; **206** Olesia Bilkei/Shutterstock.com; **208** © iStock.com/Imgorthand; **209** © iStock.com/Tomwang112; **210**r Eric Isselee/Shutterstock.com; **211**t RUBBERBALL IMAGES/NICOLE HILL/Media Bakery; **211**b Ingrid Balabanova/Shutterstock.com; **212** BJU Press; **213** © iStock.com/LouisDiBacco

Chapter 12

214–215 © iStock.com/mesteban75; **216**t Carolyn Franks/Shutterstock.com; **216**b karamysh/Shutterstock.com; **217**t FOOD COLLECTION/Media Bakery; **217**b pixelheadphoto digitalskillet/Shutterstock.com; **218**t © iStock.com/sasar; **218**c © iStock.com/psisa; **218**b JCREATION/Shutterstock.com; **219** Robert Elias/Shutterstock.com; **220** both, **227** © iStock.com/FatCamera; **221**

© iStock.com/ryasick; **222**t Westend61/Jaak Nilson/Media Bakery; **222**b Chuck Eckert /Alamy Stock Photo; **223**t Eddie J. Rodriquez /Shutterstock.com; **223**c fivespots /Shutterstock.com; **223**b H. Mark Weidman Photography/Alamy Stock Photo; **224** BJU Photo Services; **224**i © iStock.com/dszc; **225**t Carolina K. Smith MD/Shutterstock.com; **225**ct OgnjenO/Shutterstock.com; **225**cb Shannon Alexander/Shutterstock.com; **225**b © iStock .com/kevinjeon00; **229** © iStock.com/wragg; **231** Andrey_Popov/Shutterstock.com; **232** © iStock.com/Steve Debenport; **233**t Goran Bogicevic/Shutterstock.com; **233**c Syda Productions/Shutterstock.com; **233**b Romolo Tavani/Shutterstock.com; **234** robert_s /Shutterstock.com

Glossary

237t Apollo 14 Shepard/Nasa/Wikimedia Commons/Public Domain; **237**c BJU Press; **237**b Sebastian Kaulitzki/Shutterstock.com; **239**t michaeljung/Shutterstock.com; **239**c, **240**t Littlekidmoment/Shutterstock.com; **239**b Francescomoufotografo/Shutterstock.com; **240**c Anton Petrus/Shutterstock.com; **240**b Lambros Kazan/Shutterstock.com; **241**t © iStock.com/ kyoshino; **241**b u3d/Shutterstock.com; **243**t © iStock.com/Ivantsov; **243**b, **247**t © piai - Fotolia.com; **244** iStockphoto/Thinkstock; **245**t © iStock.com/SolStock; **245**b © iStock.com /cunfek; **246** both BJU Photo Services; **247**c Elnur/Shutterstock.com; **247**b Imageman /Shutterstock.com; **248**t © iStock.com/ra-photos; **248**b Getty Images/Hemera/Thinkstock; **249**t Tetra Images/Media Bakery; **249**b PaulPaladin /Shutterstock.com; **250**t VladisChern /Shutterstock.com; **250**b BIGANDT.COM /Shutterstock.com